ESSENTIALS OF POULTRY SCIENCE

MANAGEMENT, NUTRITION AND HEALTH

The Editors

Dr. Sikhamoni Haloi is a dedicated academic and researcher in animal nutrition, having completed her PhD, postgraduate, and undergraduate studies at the College of Veterinary Science, Assam Agricultural University, Khanapara. Currently, she serves as an Assistant Professor in the Department of Animal Nutrition at IVSAH, SOADU, Bhubaneshwar. Her expertise in animal nutrition is reflected in her contributions to various publications, where she integrates research findings with practical applications to enhance livestock health and productivity.

Dr. Salur Srihitha is an Assistant Professor in the Department of Animal Nutrition at the Institute of Veterinary and Animal Husbandry, SOA DU, Bhubaneswar. She completed her post-graduation in Animal Nutrition from the College of Veterinary and Animal Sciences, Mannuthy, Kerala, in 2022, and her undergraduate studies at PV Narasimha Rao Veterinary University, Hyderabad, in 2020. Along with these academic milestones, Dr. Srihitha also holds a Post Graduate Diploma in Animal Welfare from IGNOU. With a strong passion for research, she is actively engaged in advancing the field of animal nutrition.

Dr M. Bindu Pushpa is a dedicated professional in the field of Veterinary Public Health, recognized for her academic excellence and commitment to advancing knowledge in her discipline. In 2023, she completed her postgraduate degree in Veterinary Public Health from Kerala Veterinary and Animal Science University. Prior to her master's, she completed her undergraduate degree at Sri P.V. Narsimha Rao Telangana State Veterinary University in 2020. Currently, Dr Pushpa is an Assistant Professor at the Institute of Veterinary Science & Animal Husbandry, SOA (Deemed to be University), Odisha.

Dr. Uppari Anusha Rao is a dedicated veterinary professional specializing in parasitology, having earned her Master's degree in Veterinary Parasitology from KVAFSU in 2023. Prior to her master's, she completed her undergraduate degree at Sri P.V. Narsimha Rao Telangana State Veterinary University in 2020.Currently serving as a Veterinary Assistant Surgeon, she applies her clinical skills to enhance animal health and welfare. Previously, she was an Assistant Professor at the Institute of Veterinary Science and Animal Husbandry, SOA (Deemed to be University), where she taught and conducted disease diagnosis.

ESSENTIALS OF POULTRY SCIENCE

MANAGEMENT, NUTRITION AND HEALTH

Dr. Sikhamoni Haloi

Dr. Salur Srihitha

Dr M. Bindu Pushpa

Dr. Uppari Anusha Rao

ASTRAL

Publishing Authors Across the Globe

2025

Daya Publishing House®

A Division of

Astral International Pvt. Ltd.

New Delhi – 110 002

ISBN: 978-93-5919-333-5

Published by : **Daya Publishing House®**
A Division of
Astral International Pvt. Ltd.
– ISO 9001:2015 Certified Company –
4736/23, Ansari Road, Darya Ganj
New Delhi-110 002
Ph. 011-43549197, 8130496929
E-mail: info@astralint.com
Website: www.astralint.com

Printed at : **Replika Press Pvt. Ltd.**

ACKNOWLEDGEMENT

We would like to express our heartfelt gratitude to Dr. N.C. Behura, Head of the Department of Animal Nutrition, for his invaluable guidance and support throughout the development of this book. We are equally thankful to Dr. Prajna Mohapatra and all other faculty members of the Department of Animal Nutrition. My sincere thanks go to Dean Bramhadev Pattanaik for his leadership and unwavering support. I also extend my appreciation to the Institute of Veterinary Sciences and Animal Husbandry, SOA Deemed University, for providing a conducive environment for research and learning. Without the collective contributions of these esteemed individuals and the institution, this work would not have been possible.

Dr. Sikhamoni Haloi

Dr. Salur Srihitha

Dr M. Bindu Pushpa

Dr. Uppari Anusha Rao

PREFACE

Any domesticated bird species raised for their eggs, meat, or feathers is referred to as poultry. Examples of these species include chicken, ducks, quails, geese, turkeys, pheasants, and others. According to the 20th Livestock Census-2019, there were 851.8 million poultry, up 16.8% from the 19th Livestock Census-2012. Poultry is the fastest-growing sector of the agricultural industry. With an annual growth rate of less than 6.77% in 2018–19, India ranked third in the world for egg production. By 2022–23 (BAHS 2023), the country's egg production reached 138.38 billion, with 101 eggs available annually per person, compared to the ICMR's recommended 180 eggs.

India grew at a rate of about 6.77% each year between 2021–2022 and 2022–2023, producing 9.77 MT of poultry meat in 2022–2023, placing it eighth in the world. In 2022–2023, the per capita availability of poultry meat was 7.10 kg. Since its inception, the Indian poultry industry has focused mostly on chickens. Poultry, for example, offers more food at a lower cost per kg. Broiler body weight is obtained from 2 kg of feed, and the same is true for dozen eggs come from 2.2 kg. Feed in poultry yield prompt returns (e.g., one kilogram of broiler meat). is easily attained in four weeks, and at 20 weeks of age, egg production is possible. With a technology-driven economy, India has become one of the most independent developing nations in the world. It is able to produce vaccines, medications, specific pathogen-free eggs (SPF), automation systems for farms and hatcheries, pelleted feed, egg processing, poultry processing, and a national network of disease diagnostic laboratories that is accessible to both the public and private sectors.

India's agricultural economy heavily depends on the production of chicken in rural areas. Small scale chicken farmers have a greater potential to reduce unemployment, poverty, and malnutrition. From subsistence to a sustainable production system, remarkable progress has been made. The output of chicken eggs and meat must be multiplied to satisfy the increasing demand, which is one of the two main issues in poultry farming. This is significant since it is difficult to meet the current recommendations for meat and egg intake, especially in rural areas of the nation. Moreover, since cereal is the primary component of chicken feed, there will be competition for it between the human population and the poultry sector. Aside from these cereals' availability, the cost of these and other ingredients is consistently going up. Because the cost of feed accounts for over

70% of the total cost of producing chickens, finding alternate energy sources and feed ingredients becomes necessary as feed availability and pricing become more important.

The current book, "**Essentials of Poultry Science: Management, Nutrition and Health**" is a step toward more advance studies and developments in India's poultry farming industry. This book is an excellent resource for veterinary practitioners, students studying veterinary science and animal science, and progressive animal farmers.

Dr. Sikhamoni Haloi

Dr. Salur Srihitha

Dr M. Bindu Pushpa

Dr. Uppari Anusha Rao

CONTENTS

CHAPTER 1

INTRODUCTION TO POULTRY

[1]Vanlalhmangaihsanga and [1*]Nanda Kumar Roy

[1] Assistant Professor, Department of Livestock Production Management}
Institute of Veterinary Science & Animal Husbandry, SOADU, Bhubaneswar, Odisha
*Corresponding author-kumarroynanda@gmail.com

Abstract

Poultry can be defined as the rearing and breeding of domesticated fowls such as chicken, geese, ducks, turkey and some varieties of pigeon for their meat and eggs. Chickens, ducks, turkeys, and geese are of primary commercial importance, while guinea fowl and squabs are chiefly of local interest. Based on the number of animals, poultry represent the largest domestic animal stock in the world, and poultry meat was the fastest growing component of global meat production in the early 21st century. Poultry meat and eggs provide affordable high-quality protein. Poultry farming, especially on a small scale, is renewable and efficient and can provide a ready source of income and nutrition. Poultry is one of the fastest growing food industries in the world. Poultry meat accounts for about 27% of the total meat consumed worldwide and its consumption is growing at an average of 5% annually. Poultry industry in India is relatively a new agricultural industry. Till 1950, it was considered a back yard profession in India. Specific poultry stocks for egg and broilers production are available.

Keywords: Poultry, Population, Breeds, egg production, chicken production

Introduction

The term Livestock systems encompasses every aspect of the supply and utilisation of livestock commodities, such as the distribution and quantity of livestock, the various systems of production through which they are raised, projections of current and future production and consumption, the individuals involved in livestock production, and the advantages and consequences of owning livestock. In simple terms, livestock are domesticated animals raised for various products and services, including traction, meat, milk, eggs, skins, fibres, and feathers. Livestock farming can vary from intensive systems, where animals are kept in controlled environments and provided with specific diets, to extensive systems, where animals graze on natural pastures. The management and care of livestock involve ensuring their health, proper nutrition, and humane treatment. Livestock plays a critical role in many cultures and economies, contributing to livelihoods, food security, and agricultural sustainability.

Poultry farming is of great importance in the agriculture sector of India in terms of the whole picture of healthy growth and food security which not only is a means of livelihood for the farmers but also guarantees food security for the nation. This analysis is a collection of data on various sources along a timeline that constitutes a comprehensive poultry population chart of India between the 20th census, the Ministry of Fisheries, Animal Husbandry & Dairying, and the Basic Animal Husbandry Statistics 2023.

Poultry Population Statistics

Global Poultry Population

The global poultry population, dominated by chickens, ducks, turkeys, and geese, continues to expand, driven by rising demand for poultry meat and eggs. According to the Food and Agriculture Organization (FAO), poultry meat production has shown consistent growth, with chickens accounting for the largest share. The increase in poultry production is attributed to several factors, including population growth, urbanization, rising incomes, and changing dietary preferences towards lean protein sources.

Asia holds the largest share of the global poultry population, with countries like China, India, Indonesia, and Thailand being major producers. The region's growth is fueled by robust domestic demand and expanding export markets. Significant increases in chicken production were recorded in Pakistan and Vietnam. The data indicates that the global chicken population increased by 2.0% over the same period. The latter reflects the trend for pork producers to shift to poultry where possible as a result of ASF (African Swine Fever).

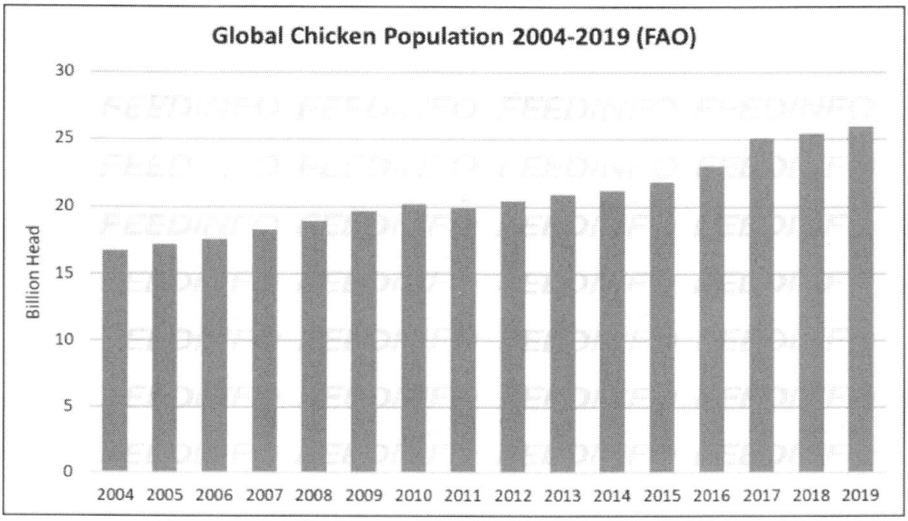

Fig 1. Global Chicken Population 2004-2019 (FAO) (Source: FEEDINFO, 2021)

Total India Poultry Population

It was mentioned based on the 20th Livestock Census, in the year 2019, the country's total poultry population has hit a record of 851.81 million, this increase in the absolute number of poultry population in the distribution of 16.81 percent can be seen from the previous Livestock Census which was conducted in the year of 2012 with a poultry population of 729.21 million.

Backyard and Commercial Poultry

Poultry farming in India influences a wide array from the traditional type of raising poultry in the backyard to the high-tech and big duck, chicken, and other poultry undertakings. This branch is not only the most energy-intensive aspect of the agricultural domain, but it also acts as an essential driving force in sustaining food safety and household income, notably in the hinterlands. This overture accomplishes the objective of all backyard and commercial poultry in India, which involves their development, significance, challenges, and prospects for the future. Competition in the poultry industry is correlated with the big increase of production in the backyard and commercial poultry sector:

Backyard Poultry

Backyard poultry is a kind of poultry farming that has been used in Indian local societies for ages and it is commonly presented by small flocks of birds in backyards. It is the backyard type of poultry that has established itself as a worthy example in rural areas and semi-urban ones, serving as a main source of income, nutrition, and social security to various households.

Growth and Significance

The production of poultry in the backyard was far more prevalent in comparison to ten years ago. As it's evident from the 20th Livestock Census, backyard poultry has touched new highs with an addition of almost 45.78% in that time from 217.49 million in 2012 to 317.07 million in 2019. This percentage of poultry backyard increase was 45.78%. This exceptional rate might be imputed to different causes, bringing in the administration's incentives, information programs, and backyard farms intrinsic merits.

Commercial Poultry

Commercial poultry farming in India today is a major agricultural sector with various livestock industries and is one of the most important contributing sub-sectors to the economy that has developed into a structured and lively organization providing huge benefits to the nation's agricultural industry. It replaces the small-scale traditional farming based on the use of local breeds of birds that are more efficient in the consumption of feed, thus making the supply chains operate at higher speeds. Consequently, the main participant of the poultry industry in India is the corporate sector that is profitable and successful in managing forums while the number of investment creators in the sector is far less in comparison.

Growth and Significance

In the last period of time, the commercial poultry industry has gained much momentum in the backdrop of an almost 4.50% increase in commercial poultry production from 511.72 million in 2012 to 534.74 million in 2019. This situation has risen with the total demand for poultry products driven by the availability of purchasing power settlements, technological development, urbanization, and controlling factor satisfaction of the people living here becoming irresistible and the price mechanism for food commodities cannot be separated from these trends.

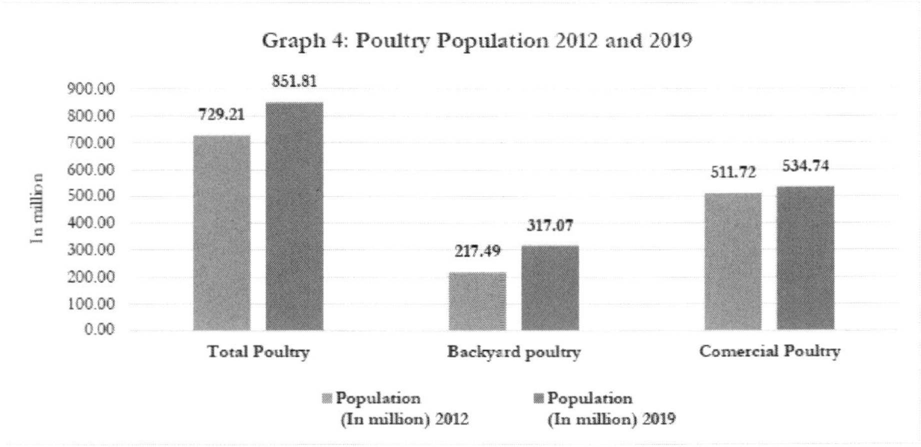

Fig 2. 20th Livestock Census Poultry Population of 2012 & 2019 (Source: 20th Livestock Census)

State-wise Poultry Distribution

At present, the proper distribution of poultry among various areas all across the country is reported as being at a high level. It is only wise to capture the details on the Livestock Census which remains the chief mechanism to report on the various real pro-County Farms among the states. This is the transport system that is inseparably influenced by local agricultural practices, climatic conditions, and market demands from the livestock production of different areas of the country. The role of the above-mentioned three states, among others, in the commercial poultry sector is of immense significance. States such as Tamil Nadu, Andhra Pradesh, and West Bengal are counted among those states that have the lion's share of the poultry population in India. The share of the country's overall poultry stock in various states varies from as low as 0,3% to as high as full coverage in some states. In practical terms, this means that poultry flocks have a wider dispersal within states influenced, however, by geographic factors and type of livestock raised.

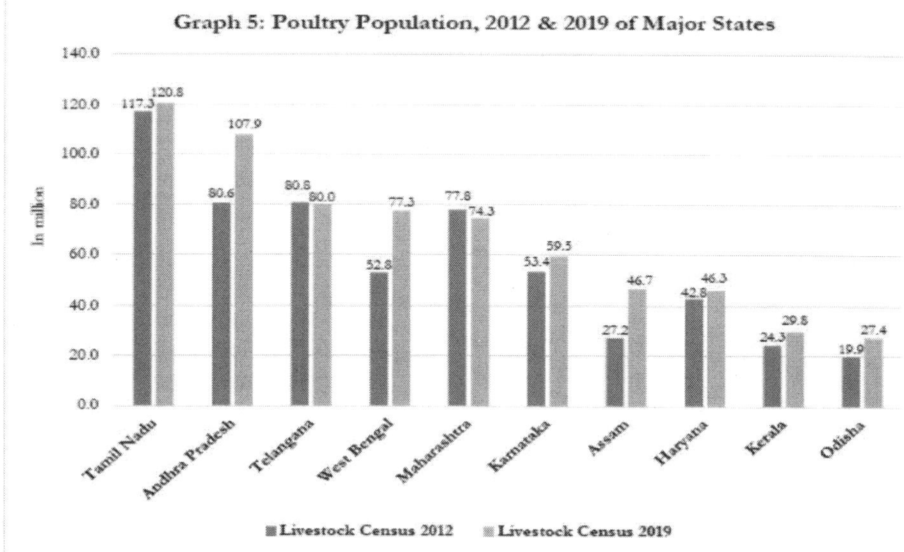

Poultry Population, 2012 & 2019 of Major States

Fig 3. Poultry Population, 2012 and 2019 of Major State (Source: 20th Livestock Census)

Breed-wise Poultry Population

Poultry farming is a significant aspect of livestock agriculture, contributing to both rural and commercial economies. Poultry is primarily categorized into four groups: Fowls, Ducks, Turkeys, and Others. These categories are further divided into Backyard Poultry and Poultry Farms. In the 20th Livestock Census, birds falling under these broad categories are bifurcated into respective breeds as recognized by the National Bureau of Animal Genetic Resources (NBAGR).

Indigenous Fowl

The Fowl category is a high-yielding segment within poultry. Indigenous fowl are categorized into Desi (native) and Improved breeds. According to the 20th Livestock Census, fowls contribute nearly 94.84% of the total poultry population.

Table 1. Breed-wise population of fowls under Backyard Poultry and Poultry Farms

Breed Name	Backyard Poultry	Poultry Farm	Total	Share (%)
Miri	54,03,648	7,445	54,11,093	0.7
Daothigir	36,94,955	1,302	36,96,257	0.5
Kadaknath	15,99,417	4,38,058	20,37,475	0.3
Punjab Brown	1,68,859	11,96,941	13,65,800	0.2
Ankaleshwar	7,02,147	6,19,859	13,22,006	0.2
Chittagong	23,855	12,17,145	12,41,000	0.2
Busra	9,95,386	2,23,338	12,18,724	0.2
Ghagus	4,55,230	3,05,734	7,60,964	0.1
Kashmir Favorolla	5,97,619	7	5,97,626	0.1
Harringhata Black	4,54,979	781	4,55,760	0.1
Danki	1,07,648	60,684	1,68,332	-
Tellichery	1,47,438	1,267	1,48,705	-
Kaunayen	1,10,706	4,122	1,14,828	-
Mewari	1,10,041	4,583	1,14,624	-
Hasli	85,733	5,652	91,385	-
Kalahasthi	11,736	38,000	49,736	-
Nicobari	25,641	974	26,615	-
Total Desi Fowl	4,08,32,670	1,16,68,843	5,25,01,513	6.5
Other Desi Fowl	18,67,31,428	5,31,78,591	23,99,10,019	29.7
Improved Fowl	5,32,09,966	46,22,31,485	51,54,41,451	63.8
Total Fowl	28,07,74,064	52,70,78,919	80,78,52,983	

As shown in the table, Improved Fowl has the highest share with 63.8%, compared to 29.7% of Other Desi Fowl and 6.5% of Desi Fowls. Among the Desi Fowl population, Other Desi Fowl have the highest contribution at 82%, followed by Aseel (12%), Miri (2%), Daothigir (1%), and Kadaknath (1%). All other recognized breeds contribute 2% of the total fowl population.

Duck Population

Indigenous Duck

The Duck category can be divided into Desi and Improved breeds. Ducks contribute nearly 3.93% of the total poultry population.

Table 2. Number of Desi and Improved Ducks under Backyard Poultry and Poultry Farms

Breed Name	Backyard Poultry	Poultry Farm	Total	Share (%)
Desi Duck	2,90,12,588	4,68,970	2,94,81,558	88.0
Improved Duck	34,89,169	5,39,116	40,28,285	12.0
Total Duck	3,25,01,757	10,08,086	3,35,09,843	

Desi Ducks hold a significant majority with 88.0%, while Improved Ducks constitute 12.0% of the total duck population.

Drakes, Ducks, and Ducklings

Table 3. The distribution of Drakes, Ducks, and Ducklings

Category	Backyard Poultry	Poultry Farm	Total	Share (%)
Drakes	77,47,641	1,38,107	78,85,748	23.5
Ducks	1,35,08,785	2,75,103	1,37,83,888	41.1
Ducklings	1,12,45,331	5,94,876	1,18,40,207	35.3
Total Ducks	3,25,01,757	10,08,086	3,35,09,843	

Ducks account for the highest share at 41.1%, followed by Ducklings at 35.3% and Drakes at 23.5%.

Turkey Population

Turkeys contribute a small fraction (0.05%) to the total poultry population.

Table 4. The distribution of Male and Female Turkeys

Category	Backyard Poultry	Poultry Farm	Total	Share (%)
Male Turkey	1,60,574	5,299	1,65,873	37.0
Female Turkey	2,62,651	19,698	2,82,349	63.0
Total Turkey	4,23,225	24,997	4,48,222	

Female Turkeys have the highest share at 63.0%, compared to 37.0% for Male Turkey.

Other Poultry Birds

The category of Other Poultry Birds, which includes Quails and other species, contributes nearly 1.17% to the total poultry population.

Table 5. The distribution of other categories of poultry

Category	Backyard Poultry	Poultry Farm	Total	Share (%)
Quails	19,32,270	48,86,382	68,18,652	68.5
Other Poultry Birds	14,02,737	17,33,663	31,36,400	31.5
Total Others Poultry	33,35,007	66,20,045	99,55,052	

Quails dominate this category with a 68.5% share, while other poultry birds constitute 31.5%.

Egg and Meat Production

Egg Production

Egg production in India has shown remarkable growth over the years, making the country a significant player in the global poultry industry. According to the "Basic Animal Husbandry Statistics - 2023," India's total egg production for the year 2022-23 was 138.38 billion eggs, positioning India as the third-largest egg producer globally, according to the FAO. This section provides an in-depth look into the trends, state-wise contributions, and factors influencing egg production in India.

The total egg production in India has been on a consistent rise, increasing by 6.77% from the previous year (2021-22). This growth trend is reflected in the increase in per capita availability of eggs, which reached 101 eggs per annum in 2022-23, up from 95 eggs per annum in 2021-22. This steady growth indicates improvements in poultry farming practices, enhanced productivity, and increased consumer demand for eggs as a nutritious food source.

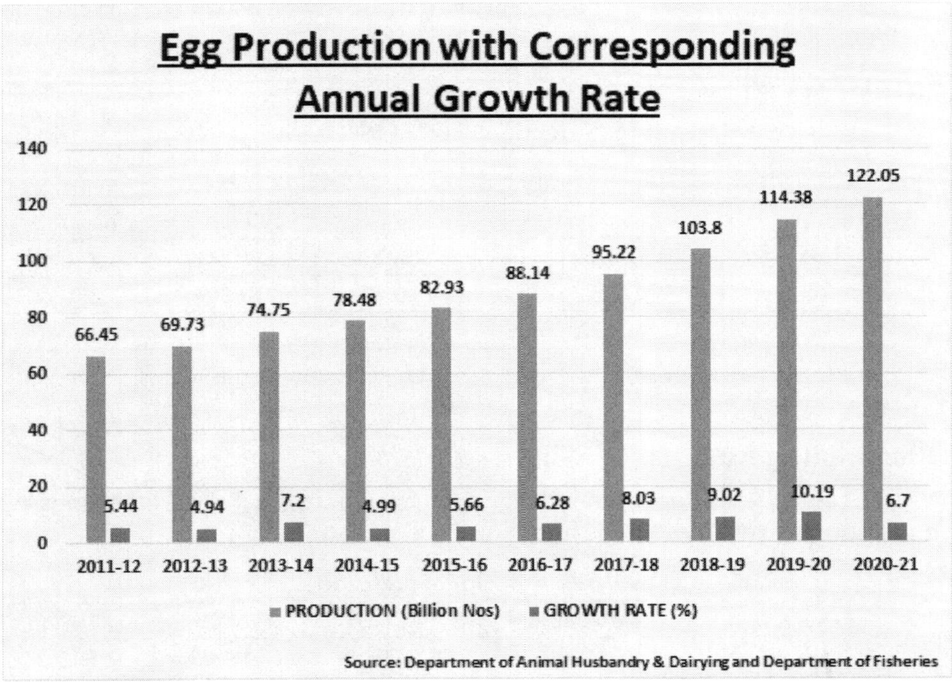

Fig 4. Egg Production with Annual Growth rate, 2011-2021

Species-wise Contribution

The species-wise breakdown of egg production in 2022-23 reveals that improved fowls are the predominant contributors, accounting for 88.41% of the total egg production. Desi fowls contribute 10.73%, while desi ducks and improved ducks contribute 0.74% and 0.12%, respectively. This distribution highlights the dominance of commercial poultry farming in India, with improved breeds playing a crucial role in meeting the high demand for eggs.

State-wise Egg Production

India's egg production is concentrated in a few key states, with Andhra Pradesh, Tamil Nadu, and Telangana leading the way. These three states together account for nearly half of the country's total egg production.

- **Andhra Pradesh**: The largest producer, contributing 20.41% of the total egg production.
- **Tamil Nadu**: Second largest, with a share of 16.08%.
- **Telangana**: Third, contributing 12.86%.

Other significant contributors include West Bengal (8.84%) and Karnataka (6.38%). These five states together produce 64.93% of the total eggs in India, underscoring their critical role in the country's poultry industry.

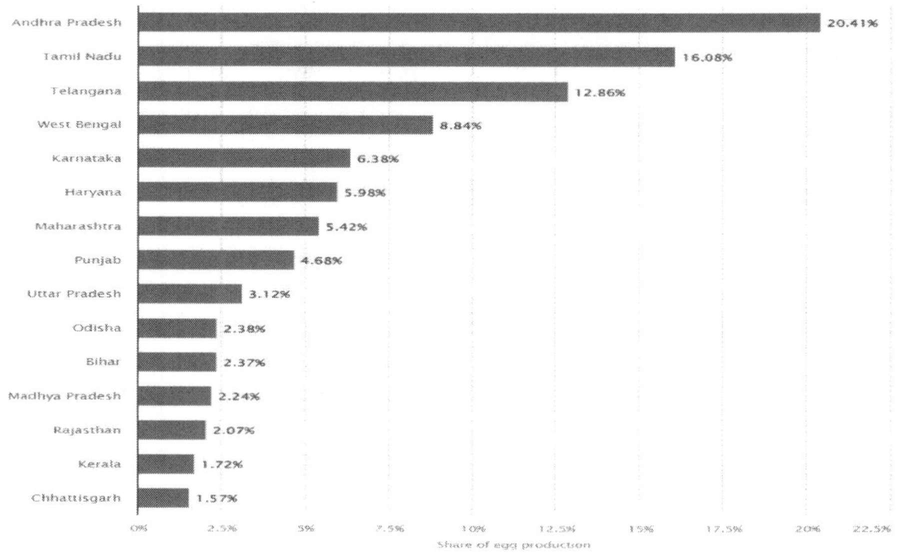

Fig 5. Distribution of egg production in India during financial year 2022 (Source:www.statista.com/india-egg-production-major-state)

Annual Growth Rates by State

The annual growth rate of egg production varies significantly across different states. The national growth rate for egg production in 2022-23 was 6.77%. Several states exceeded this average, with notable growth in West Bengal, Sikkim, Uttar Pradesh, Odisha, Madhya Pradesh, Karnataka, Chhattisgarh, and Bihar. These states have shown substantial improvements in their poultry farming practices, contributing to the overall increase in egg production.

Per Capita Availability of Eggs

The per capita availability of eggs is an important indicator of both production efficiency and consumption patterns. In 2022-23, nine states had a per capita availability higher than the national average of 101 eggs per annum. This reflects the regional disparities in egg production and consumption, influenced by factors such as dietary preferences, economic conditions, and local farming practices.

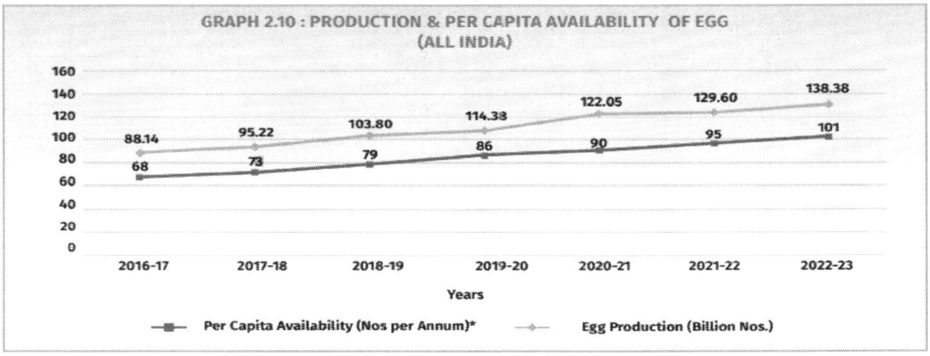

Fig 6. Production & per capita availability of egg (Source: Basic Animal Husbandry Statistics – 2023)

Meat Production

Poultry meat production in India has emerged as a vital component of the country's meat industry, contributing significantly to the overall meat production. In the year 2022-23, poultry meat production continued to dominate, reflecting the sector's growth, technological advancements, and increasing consumer demand for poultry products. This section provides a comprehensive overview of poultry meat production in India, highlighting its contribution, growth trends, and factors influencing its success. The diagram shows that nearly 51.14% of meat production is contributed by Poultry while Buffalo, Goat, Sheep, Pig, and Cattle contributes nearly 17.61%, 14.47%, 10.51%, 3.85% and 2.43% of meat production respectively to the total meat production of the country

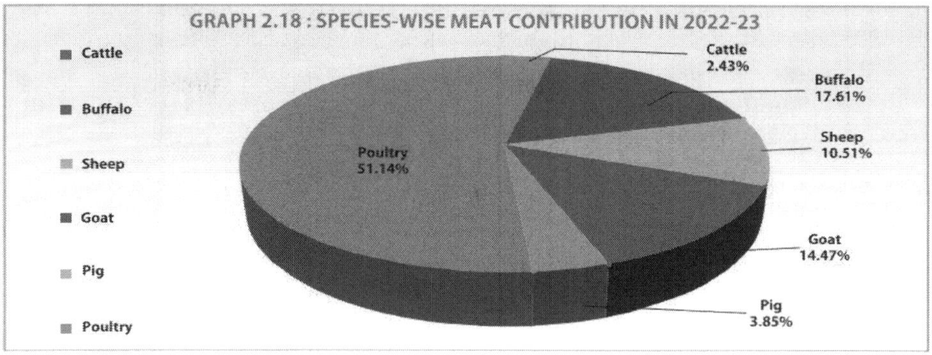

Fig 7. Species-Wise Meat Contribution, 2022-23 (Source: Basic Animal Husbandry Statistics – 2023)

Growth Trends in Poultry Meat Production

The growth in poultry meat production has been remarkable over the years. The total meat production in India increased from 7.39 million tonnes in 2016-17 to 9.77 million tonnes in 2022-23, with poultry being the largest contributor. This growth trend indicates a consistent increase in poultry meat production, driven by advancements in farming practices, genetic improvements in poultry breeds, and better disease management.

The annual growth rate for meat production, including poultry, was 5.13% in 2022-23. While this figure represents the overall meat production, poultry meat's substantial contribution suggests that the sector has experienced similar or higher growth rates, reinforcing its leading position in the industry.

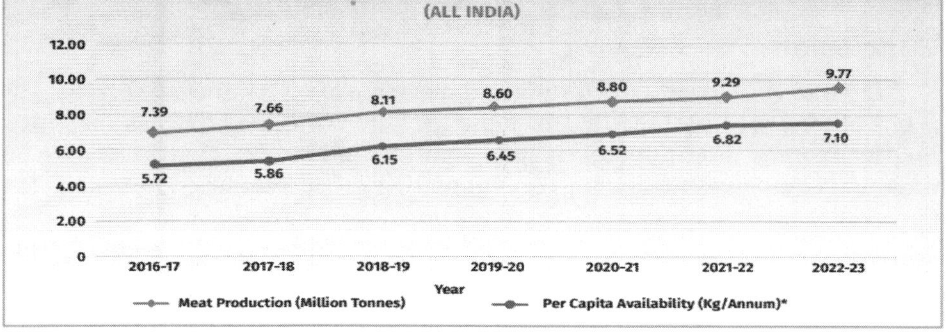

Fig 8. Production & per capita availability of meat (Source: Basic Animal Husbandry Statistics – 2023)

The total Meat production in the country was at 9.77 Million Tonnes in the year 2022-23. The per capita availability was 5.72 kg per annum during the period 2016-17. There has been steady increase in per capita availability of Meat. The per capita availability reached at 7.10 Kg/Annum the year 2022-23, a 0.28-point jump from 6.82 Kg/Annum of the previous year 2021-22.

Regional Contributions to Poultry Meat Production

Poultry meat production is concentrated in certain key states, which are major contributors to the country's total meat output. States like Andhra Pradesh, Tamil Nadu, and Telangana are leading producers, thanks to their well-developed poultry farming infrastructure and favourable climatic conditions. These states have invested significantly in modern poultry farming techniques and have large-scale commercial poultry farms that cater to both domestic and international markets.

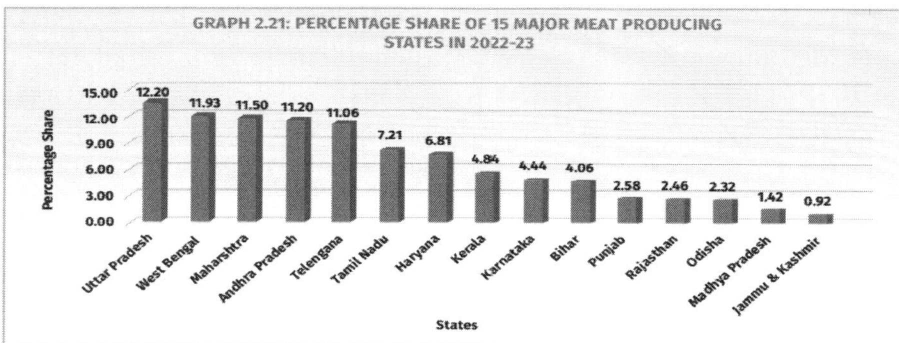

Fig 9. Percentage share of 15 major meat producing (Source: Basic Animal Husbandry Statistics – 2023)

Conclusion

The main poultry products, Egg and meat, are of high nutritional value. This is one of the significant reasons for rapid growth in the need for these products. Poultry farming is a profitable business with a continuous source of income. As it requires less investment compared to rearing other livestock, any small-scale farmers can also run this practice as a side business. Poultry is capable of utilizing wastes like bran, substandard grains, vegetables which normally go as waste by feeding on them. This way the farmer can save a portion of his expenses towards poultry feed. A properly maintained shelter, nutritious feed and disease prevention measures can aid good quality and quantity of products.

References

Agricultural Statistics at a Glance 2021 (1st ed., p. 431). Government of India Ministry of Agriculture & Farmers Welfare Department of Agriculture & Farmers Welfare Directorate of Economics & Statistics. www.agricoop.nic.in & https://desagri.gov.in

2019 Global Livestock Population – FAO Data. FAOSTAT. https://www.fao.org/statistics/en/

Annual Report 2021-2022 (pp. 141-144). Department of Animal Husbandry and Dairying Ministry of Fisheries, Animal Husbandry and Dairying Government of India.

Basic Animal Husbandry Statistics - 2023 (pp. 15-22). Government of India Ministry of Fisheries Animal Husbandry and Dairying Ministry of Animal Husbandry and Dairying Krishi Bhawan, New Delhi.

Breed-wise report of livestock and poultry Based on 20th Livestock Census (pp. 67-75). Government of India Ministry of Fisheries Animal Husbandry and Dairying Ministry of Animal Husbandry and Dairying Krishi Bhawan, New Delhi. https://dahd.nic.in/ahs-division/20th-livestock-census-2019-all-india-report.

Naik, P. K., Swain, B. K., & Beura, C. K. (2022). Duck production in India - A review. Indian Journal of Animal Sciences, 92(8), 917-926.

https://doi.org/10.56093/ijans.v92i8.118712.

Jairath, G., Gopinath, D., Rialch, A., Singh, P., Radotra, S., & Bishnoi, S. (2022, June 14). Overview of the Indian Meat Industry. SR PUBLICATIONS. Retrieved August 3, 2024, from https://www.srpublication.com/overview-of-the-indian-meat-industry/#:~:text=Half

CHAPTER 2

HOUSING MANAGEMENT OF POULTRY

*[1]Nanda Kumar Roy and [2]Sikhamoni Haloi

[1] Assistant Professor, Department of Livestock Production Management, Institute of Veterinary Science & Animal Husbandry, SOADU, Bhubaneswar, Odisha
[2] Assistant Professor, Department of Animal Nutrition, Institute of Veterinary Science & Animal Husbandry, SOADU, Bhubaneswar, Odisha
*Corresponding author- kumarroynanda@gmail.com

Abstract

Housing and equipment's cost accounts for most of the fixed capital of a poultry farm. Housing refers to keeping the poultry birds under a structure where they are safeguarded and comforted from troubles (Physical, Psychological or Physiological) and the owner can organize the routine work most effectively for optimum return.An unthoughtful expensive house with poor efficiency might became a deciding factor for successful and failure of a poultry farm. A well-designed poultry farm will pay for itself through better production, saving of labour and avoiding disease incidence in long run. Basically, the poultry house should be designed to protect birds from extreme weathers, parasites & predators to avoid the development and spread of infectious disease: to enable the poultry attendant to work efficiently without wasting time in unnecessary, movements.

Keywords: Poultry, housing systems, housing management

Introduction

Poultry production in India is a significant element of the country's agricultural and food sectors. Poultry is one of the fastest growing parts of the agricultural sector. The composite periodic growth rate of poultry protein between 2015 and

2025 is estimated to be +2.4%. Worldwide in the tropical regions extreme weather have proven generally detrimental to livestock production and is particularly of interest in poultry sector because of the latter's high sensitivity to temperature change. Just like mammals, the avian species have the capability to regulate their body temperatures by losing or generating heat in response to environmental temperature. If the body temperature of a bird, which normally runs between 39.4 and 40°C, is allowed to increase, the bird will not perform well. Heat stress in poultry is one of the major challenges in poultry production which causes nutrition-deficient, poor growth, reduce egg production and egg size, and at last high mortality of the birds. This problem is further compounded by the high body heat generated by genetically improved variety laying birds with increased metabolic activity resulting from the high rate of egg production.

Poultry housing plays a vital part in the determination of the internal climatic conditions of the house for optimum health, growth and productive performance of the birds. Housing is crucial for poultry for several reasons, all of which impact their health, productivity, and overall well-being. Here's why proper poultry housing is so important:

1. **Protection from Weather & Predators**: Poultry housing shields birds from extreme weather conditions, including excessive heat, cold, rain, and predators.
2. **Disease Control**: Good housing design helps prevent the spread of diseases. Proper ventilation, sanitation, and spacing reduce the likelihood of respiratory infections, parasitic infestations, and other health issues.
3. **Nutritional Efficiency**: Well-designed housing ensures that poultry have easy access to food and water, which is crucial for their growth and egg production. Housing should be designed to minimize feed wastage and water spillage.
4. **Comfort and Welfare**: Providing adequate space, proper ventilation, and bedding material contributes to the comfort and welfare of poultry. Overcrowding and poor conditions can lead to stress, aggression, and reduced productivity.
5. **Waste Management**: Good housing systems facilitate the management of manure and waste, which helps in maintaining a clean environment and reducing the risk of disease.
6. **Ease of Management**: Properly designed poultry housing makes it easier for farmers to manage the flock, including routine tasks like feeding, cleaning, and health checks.

In summary, good poultry housing is integral to maintaining a healthy and productive flock. It provides protection, promotes good health, ensures efficient management, and supports the overall well-being of the birds.

Factors affecting for selection of location of poultry farm

1. **Location:** A commercial farm should not be located in a congested area or in close proximity to existing farms which will affect the poultry performance due to noise and dust pollution. From the nearby existing farm minimum 1 km distance should be maintained.
2. **Soil type:** The ideal soil for poultry house constructed area should be sandy loam. The perfect type of soil reduces construction expenditure, dampness, avoids seepage, erosion etc.
3. **Topography:** The particular area where the poultry house is planned that area should be elevated than the surroundings for proper drainage and clean surroundings. A relatively labelled area requires less site preparation, thus lowering building cost.
4. **Drainage:** The soil should be porous and the slope gentle so that drainage is efficient and farm premises remains dry.
5. **Sunlight exposure and wind protection:** The poultry house should be located in such a way that long axis is in east-west direction which will protect the birds from the direct sunshine. The floor of the poultry farm should expose to sunlight to remove dampness which will prevent the growth of different disease-causing agents inside the farm. Tall quick growing trees should be grown near the building to protect from wind and solar radiation.
6. **Nearby market:** Whatever is produced in poultry farm has ultimately to reach market. Hence nearer the market, lesser will be loss due to spoilage of products during storage and transport. Thus, farm should be located nearer to towns, but not in towns as this will offset some other advantages listed below.
7. **Water supply:** Water should be available cheaply and in plenty, with the availability of electricity for running tube wells.Before selection the site for construction the water quality should be tested for different impurities and contaminations.
8. **Transport facility:** There should be Good Road facility and also have the accessibility to reach the market. Good and easy transportation facility will enhance the transportation cost and also it minimizes spoilage of products.
9. **Electricity supply:** Enough and continuous supply of electricity should be there for better efficiency of labour, speedy work, better security. It is needed for operating various machines used in the farm and is the light source to the birds
10. **Availability of cheap labor:** Nearby by the location of the poultry farm, there should be availability of the experienced farm labour which can be managed with cheaper wages.

Poultry house construction

Some important factors which are very important for construction of a poultry farm are as follows:

1. **Orientation (Direction)**: Orientation of the building with respect to wind and sun consequently influence with force, precipitation, temperature and light on different external surfaces. Poultry house should be built in such a way that long axis should be in east-west direction which will prevent the direct sunlight to the birds.
2. **Size**: The size of the poultry house depends on the various factors like farmers capital investment, total number of birds will be reared in the farm etc. Usually, it is calculated on the basis of total floor space requirement of the total number of birds reared in the farm. As a thumb rule, the floor space requirement of broiler bird is 1 square foot and for layer it is 2 square feet.
3. **Length**: The length of the poultry house is affected by the total number of birds reared and the availability of land for farming. Several factors like total no of birds reared, their floor space requirement as per the category and age of birds affects the length of the poultry house. Usually, the length of the poultry house should not be exceed four to six times with the width of the house.
4. **Width**: Usually, the width of the poultry house should not be more than 25 to 30 feet in tropical countries. This is mainly due to the proper ventilation and aeration in the mid portion of the poultry house. Poultry house with more than 30 feet of width, should be installed with ridge type of ventilation at the middle line of the roof top of the shed with proper overhang. In case of environmentally controlled poultry house, the width may be 40 feet or more as the ventilation is totally controlled with the help of different types of exhaust fans.
5. **Height**: Height of the poultry shed vary as per the design and type of the poultry house. Usually, the height of the sides from foundation to the roof line should be 6 to 7 feet which is considered as eaves height and height at the centre of the shed is 10 to 12 feet.
6. **Foundation**: The foundation of the poultry shed should be strong, solid of bricks, pucca to prevent rodents, prevent seepage of water into into poultry shed and provide safety to birds. The foundation of the poultry shed should be made of concrete with 1 to 1.5 feet below the surface and 1 to 1.5 feet above the ground level.
7. **Floor**: The floor of the poultry shed may vary according to the needs, means and type of available resources. Floor may of different types like mud floor, brick floor, concrete floor, stone floor, slat floor etc. Among these the desired type of floor is the pucca or cemented which is impervious to moisture with slope towards door. The concrete floor should be extended

with 1.5 feet outside the wall on all sides of the shed to prevent rat and snake into the house.

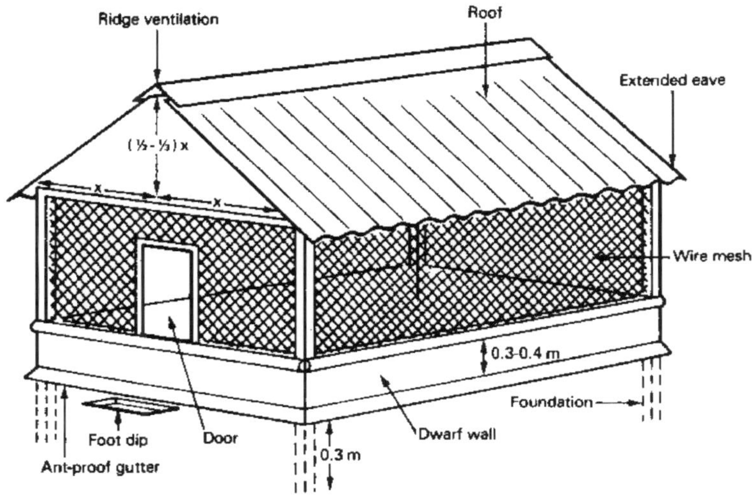

Fig 1. A Poultry house (Source: FAO agricultural services bulletin, 2003)

8. **Doors**: The door doors should be located in the end of side walls opposite to each other and covered with wire netting. In case of deep litter system, the door must be open outside.
9. The size of door is preferably 6 x 2.5 feet. A foot bath should be constructed at the entry point which will prevent the outbreak of various contagious and infectious disease into the farm from the visitors. The foot bath always should be filled with disinfectant mixed with clean water.
10. **Side walls**: The side wall is very important during the rainy seasons and also for adequate ventilation for the birds. In case of broiler shed, the height of the sidewall should be 1-1.5 feet and normally it should be at the bird's back height.
11. **Roof**: There are various types of roofing materials available for the poultry shed like Straw, plant leaves, Tiles (Country and Mangalore), Plastic, Asbestos, Aluminium, Fibre glass etc. selection of the roofing materials depends on the budgets of the farmers.

Type of roof

Shed: This type of roof is out dated. This has a wall which is totally enclosed on one side with the roof slanting forward. Ventilation is only on one side. No cross ventilation

- Gable: This is suitable for all type of houses. There is good cross ventilation. For environmentally controlled house, this type of roof is highly suitable

- Half Monitor: It is more suitable for the tropics. In this system, one roof overrides the other roof leaving a gap of 6″ which is closed by using a mesh.
- Full Monitor: This is costly and needs a special truss. It is provided with ridge ventilation in which warm air within the house escapes out and the vacuum is filled by fresh air drawn from the outside

Among different types gable type is commonly preferred in tropical countries like India.

Overhang

The overhang of the roof should not be less than 3.5 feet in order to prevent the entry of rain water into the shed.

Lighting

Light should be hanged from ceiling and it should be provided at 7-8 feet above the ground level. In case of incandescent bulbs, the interval between two bulbs is 10 feet and in case of tube lights (fluorescent lights) the interval should be 15 feet.

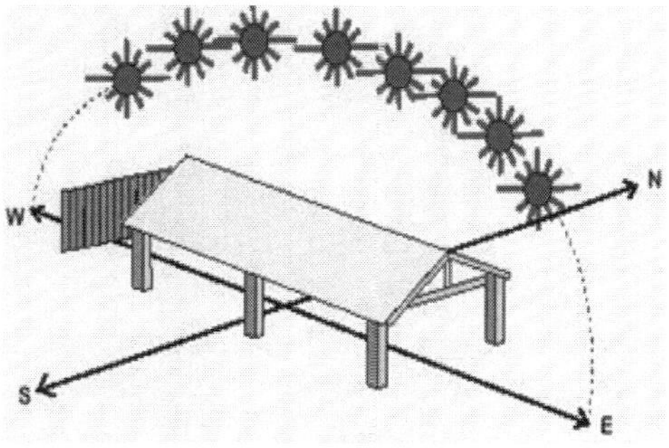

Fig 2. Direction of the poultry house

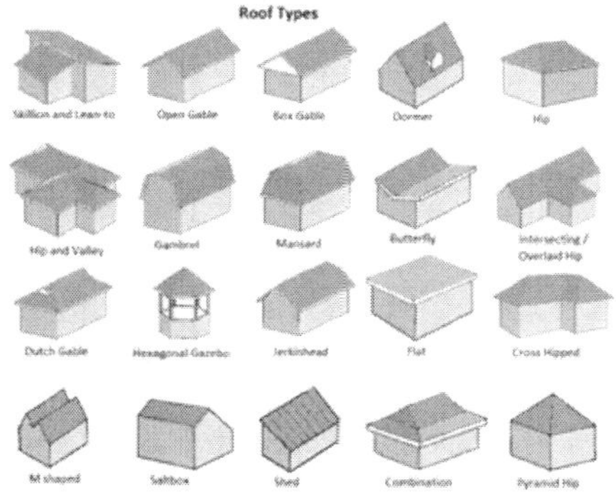

Fig 3. Different types of roofs (Source:www.asiafarming.com/poultry-farming-information-guide and https:// theconstructor-poultry-roof.org)

Different types of poultry houses

Depending upon the activity on poultry farms following buildings may be needed. If more than one type of houses is to be constructed on one farm, these buildings should be located in such a way that it cuts down the chances of cross contamination to minimum, affords maximum labour efficiency and better supervision of farm with least effort.

- **Brooder / chick house**- In this type of house usually layer birds from 0 to 8 weeks of age are reared and sufficient brooding is done in the shed.
- **Grower house**-In the grower house, the egg type birds from 9 to 18 weeks are age groups are reared.
- **Brooders cum grower house**-In this type of house, the layer birds from 0 to 18 weeks of age groups are reared. Brooding and growing is completed in this type of house.
- **Layer house**-In layer house, the layer birds which are more than 18 weeks of age are reared till 72 weeks of age.
- **Broiler house-**In this type of house broiler birds which are usually reared for the meat purposes are kept from 0 days till their marketing period (6 weeks).
- **Breeder house-**This type of house is usually made for the poultry breeder birds where both male and female breeders are maintained at appropriate sex ratio.
- **Environmentally controlled (EC)** house-Environment Control (EC) housing is the modern method of housing in which inside conditions are maintained as near as possible to the bird's optimum requirements.

System of poultry Housing

Poultry can be housed under different systems based on several factors like availability of land, cost of land, type of farming activity, climatic condition and labors' availability etc.

Broadly, poultry housing system can be divided into following types:

1. Backyard Poultry farming system
2. Free-range or extensive system
3. Semi-intensive system
4. Intensive system
5. Environment control housing system

1. Backyard Poultry farming system

Backyard poultry farming is the old poultry farming system where small numbers of local chicken birds are reared by each household with or without inputs in the backyard of the villages. Under this system the poultry birds are allowed for foraging outside provided with supplementary feeding and clean water during the day time and during the night they are provided with locally available low-cost shelter. Some varieties of chickens have been developed for rearing under this system such as Giriraja, Grampriya and swarnadhara.

2. Free range system

Free range method of poultry rearing is the oldest methods of poultry production by the framers where there is no shortage of land. In this system, the poultry birds are allowed to loose on ground which is restricted by a fence in the surroundings. Birds' full fills their daily feed requirement by foraging. During the night, shelter is provided by temporary roofing supported by ordinary poles. This system is basically preferred for organic egg production. Usually in this system we can rear about 250 adult birds per hectare of land.

3. Semi intensive system:

In this system of rearing, the birds are half-way reared in houses and half-way on range, Solid floor houses are prepared for night shelter or as per need in the day time also and also given open access to runs in the field. Scientific feeding and watering facilities are provided in the pen. On an average, the stocking density rate for adult birds is 750 per hectare.

4. Intensive system

This system is adopted where the farmers have limited land he wants to raise a large commercial farm. Under this system the poultry birds are totally confined to the house entirely, with no access to land outside. The traditional system of maintaining poultry under free range or semi-intensive has been replaced to a large extent by the intensive system. Under intensive system, large number of birds are reared under deep litter or cage system.

Advantages	Disadvantages
1) Minimum land is required for farming. 3) Day-to-day management is easier. 4) The production performance is higher 5) Scientific Poultry management practices like feeding, breeding, culling, medication, etc. can be applied easily and accurately.	1) Birds 'welfare is affected. They cannot perform the natural behaviour 2) Balanced diet should be provided to avoid nutritionally deficient diseases. 3) Chances for spreading of different diseases are comparatively more.

There are two types of intensive system namely **Deep litter and Cage system**.

Deep Litter System

Under this system the birds are kept on suitable litter material of about 3" to 5" depth inside the house till their marketing age. Different types of litter materials like paddy husk, saw dust, ground nut hulls, chopped paddy straw or wood shavings are used. At every fortnightly, the litter is spread on the floor in layers of 2″ height till the required height is obtained. Feeding and watering arrangement are made inside the house.

Advantages	Disadvantages
1. Safety of birds: The birds and eggs are safe in deep litter house.	1. Due to the direct contact between bird and litter materials, occurrence of different disease may be a problem.
2. Litter as a source of nutrient: The build-up litter supplies few nutrients to the birds.	2. Due to dust from the litter, respiratory problems may arise.
3. Disease control: The level of worm infection is much lower than the birds kept in semi-intensive method.	3. The cost of litter is an extra expenditure on the bird's production cost.
4. Labour saving: There is no need to clean a pen up to selling of broilers.	4. If there is any ventilation faults, it can have serious issues than the cage system

Fig 4. Poultry rearing in Backyard farming (Source: Practical skill son backyard poultry production in India by P Kushwaha)

Fig 5. Poultry rearing in free range system (Source: https://pamora-farm-chicken.html)

Fig 6. Poultry rearing in semi-intensive system (Source:www.youtube.com/watch?v=Tdfz5TAv7CM)

Fig 7. Broiler house under deep litter system (Source:https://litter-management-in-poultry.html)

Cage system

In this system, poultry birds are reared in a raised wire netting floor in smaller compartments, which is called as cages, either fitted with stands on floor. Feeders and waterers are attached to cages from outside. In Nipple watering system, pipeline is installed through or above cages. Worldwide at present situations 75%

of commercial layers are kept in cages. Auto-operated feeding trolleys which is connected with the feeding line from outside and inside and egg collection belts can also used in this rearing system. Depending on type of cages, the droppings are either collected in trays underneath cages or on belts or on the floor or deep pit under cages.

Advantages	Disadvantages
Minimum floor space is needed	1) High initial investment cost.
Less feed wastage & good FCR	2) Handling of manure may be problem.
Protection from internal parasites	3) Occurrence of blood spots in egg is more
Sick birds can be identified and eliminated.	4) Problem of cage layer fatigue.
Clean eggs production Less vices(egg eating, pecking)	5) In case of broilers, when the weight is more than 1.5 kg, incidence of breast blisters is more.

Types of cages

Based on the number of birds reared in a cage:	**Based on the number of rows:**
1. Single or individual bird cage (1 bird in cage) 2. Multiple bird cage (From 2 to 10 birds, usually 3 or 4 birds per cage) 3. Colony cages (Holding birds >11 per cage)	1. Single-deck 2. Double-deck 3. Triple-deck 4. Four-deck 5. Flat-deck
Based on arrangement of cages:	**Based on the type of bird reared:**
1. Stair-step cages • M-type cages • L-type cages 2. Battery cages	1. Brooder / chick cages 2. Grower cages 3. Layer cages 4. Breeder cages 5. Broiler cages

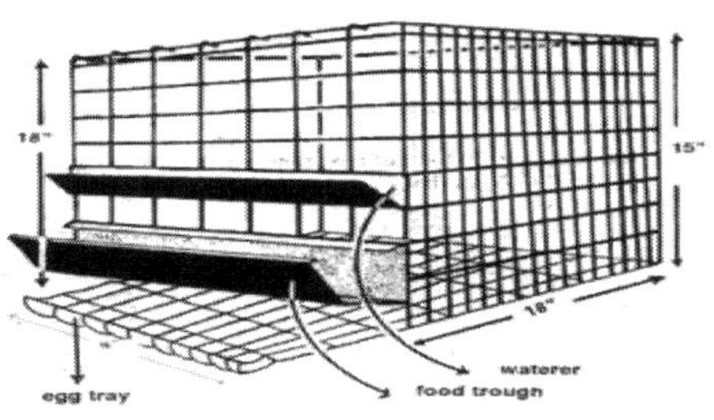

Fig 8. Conventional cage . Fig 9. Californian cages cage
(Source: TNAU Agritech Portal) (Source:www.indiamart.com/california-type-poultry)

Table 1. Floor Space requirement for different age groups of birds

Type of bird	Age(weeks)	Floor space (Sq. ft.)	Feeder space		Waterer space	
			(inch)	(cm)	(inch)	(cm)
Deep litter system						
Layer i) Chicks	0 – 4	0.50	1.00	2.50	0.50	1.25
	4 – 7	1.00	2.00	5.00	0.50	1.25
ii) Grower	8-12	1.50	3.50	8.00	0.75	2.00
	13 – 20	2.00	5.00	12.00	0.75	2.00
iii) Layer	21 – 72	2.50	6.00	15.00	1.00	2.50
Broiler	0 – 4	0.60	2.00	5.00	0.50	1.25
	4 onwards	1.20	3.50	8.00	1.00	2.50
Cage system						
Layer i) Chicks	0 – 7	0.24	1.00	2.50	0.50	1.25
ii)Grower	8 – 20	0.48	2.00	5.00	0.75	2.00
iii)Layer	21 – 72	0.72	3.00	7.50	1.00	2.50

Environment control housing

Environment Control (EC) housing is the modern method of housing in which inside conditions are maintained as near as possible to the bird's optimum requirements.In order to maintain proper temperature, ventilation, relative humidity and lighting program in the house, the shed is constructed longitudinally preferably east to west, fitted with big exhaust fans on west side and evaporative cooling pads on east side along with automatic drinking and feeding systems.

Advantages	Disadvantages
1. Less floor space requirement	5. High initial investment.
2. Better prevention &control of diseases.	6. Skilled labour is required.
3. Faster and uniform growth.	7. Requires 24 X 7electric supply.
4. Easy for management.	

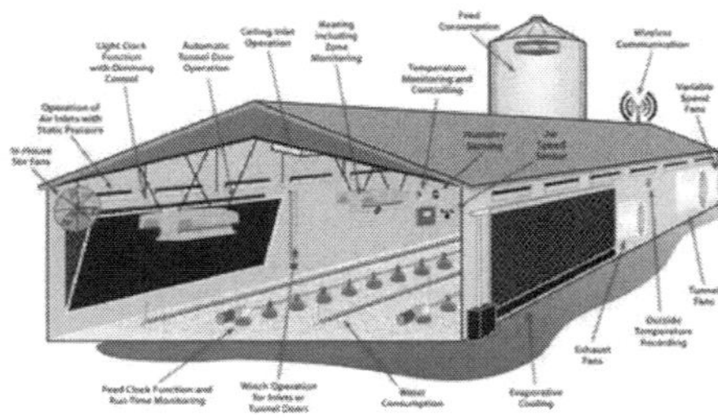

Fig 10. Environment control poultry house **Fig 11. Poultry bird reared in EC house**
(Source:www.poultry-Chicken-broiler-design-house.in)

Control over microenvironment: The control over microclimate can be achieved by completely eliminating macroclimate/ environment contact with the house which can be further achieved by completely enclosing from outside inclimental weather conditions.

Ventilation

In the poultry house, the provision of ventilation system is must for proper exchanges of oxygen and carbon dioxide and to remove dust particles and various gases from the house. Basically, artificial ventilation are two types: positive and negative ventilation. In environmentally controlled house generally negative ventilation is used.

Negative pressure ventilation: Negative pressure ventilation usually creates a partial vacuum that pulls air into the poultry house evenly through all inlets which creates more uniform conditions inside the house.

Types of Negative-Pressure Ventilation Operations:

1. **Minimum ventilation** - (also called just "power ventilation" or even "power vent")- operated on a timer and used for cooler weather and/or smaller birds.
2. **Transitional ventilation** – operated on thermostat or temperature sensor and used for heat removal when wind-chill (tunnel) cooling is not needed or desirable.
3. **Tunnel ventilation** –When it is warm outside and the ventilation system operates in tunnel mode, fresh air enters the house via air inlets that are located close to the gable. The fresh air is »pulled« through the poultry house in longitudinal direction by with the help of high-performance fans.

Components of mechanical ventilation: Mechanical ventilation systems consist of four major components: fans, openings, heaters, and controls.

1. **Fans:** In mechanical ventilating systems, fans are basically used to supply the energy needed to exchange the required amount of air in a poultry house in each minute.

2. **Openings:** The prime functions of air inlets inside the EC house are to provide fresh air throughout the shed, control the direction of airflow, and to maintain sufficient inlet air velocity. For negative-pressure ventilation air inlets in poultry housing include continuous slots and area inlets.
3. **Heaters:** During cold weather supplemental heat is usually required in naturally ventilated grower poultry houses to maintain desired indoor temperatures. For supplemental heating in poultry houses, various types of heaters are used like radiant, space, and make-up air heaters etc.
4. **Controls:** They are needed to adjust ventilating rates (fan controls), supplemental heating rates, and the air velocity rates (fan controls), supplemental heating rates, and the air velocity through openings as weather, bird age and size change.

Cooling the house: When temperature increases above 85^0 F (29.4^0 C) then for providing comfortable environment to the birds cooling system are required. Generally cooling in the poultry houses is done by four methods:

1. **Low pressure fogging system:** Fogging nozzle that operate at regular water pressure are installed throughout the house or over the birds in the cages.
2. **Pad –and- fan system:** Exhaust fans in the house draw incoming air through a wet pad where the evaporation of moisture from the pad reduces the incoming air. For cooling the shed, two types of system used. a. Pressurized system: Evaporative coolers are fitted outside the house and air is sucked through the evaporative pads of the coolers and then finally forced into the poultry sheds. b. Vacuum system: Evaporative pad is fitted at one end of the house and exhaust fan are placed at another end of sheds.
3. **Fog –and- fan system:** It is similar to pad and fan system except the incoming air is drawn through a hood in which high pressure floggers have been installed. The cooling effect is significantly increased by airflow from the use of fans within the shed.
4. **High pressure nozzle system:**. Fogging basically works best in the dry climates, and usually involves various rows of high-pressure nozzles that release a fine mist in the whole poultry house which converts the liquid phase into vapour phase.

Feeding system in environmentally controlled poultry house: For efficiently running the feeding in the whole house a control system requires some way of knowing regarding when to turn the feeders on and when to put it off. Basically, the norms are that the same amount of feed should be available at all locations along the feeder in the whole farm. Feed monitoring systems are available to measure the heaters, amount of feed consumed by bird. Generally, in environmentally controlled houses 7- 8 times per day feeding is practised.

Lighting management in environmentally controlled house: Environmentally controlled house must be lightproof; that is, no outside light should be allowed to inter in the building. For smooth operation in the 24 hours lighting programme, acontrol and monitoring system should provide scheduling of lights which is easy to use in the farm.

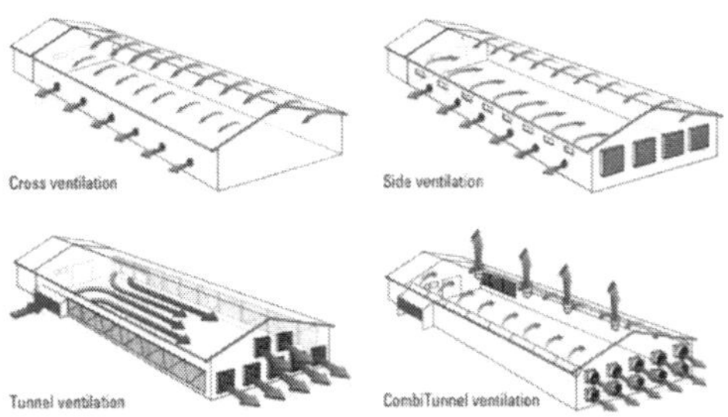

Fig 12. Different ventilation in EC house

Fig 13. Feeding and watering line in EC house

(Source: https://ventilation-system-for-a-poultry-farmhouse) (Source:www. poultrytrends.in)

Conclusion

Proper poultry housing plays a major role in the determination of the internal climatic conditions of the house for optimum growth, productive performance and health of the birds.Rear the birds under a scientific shed where they are safeguarded and comfortable from troubles and the farmer can organize the routine work most effectively for optimum benefit from the farm. In the poultry production, heat stress had resulted in stunted growth, under-nutrition, reduction in egg production and size, laying of premature eggs and in some cases death also occurs. Proper scientific housing provides comfort and protects the poultry birds from direct sun rays, wind, rain and predators. The poultry house should be constructed considering the comfort level of the birds and it should be well ventilated, warm during winter, reasonably cool in summer and free from drafts. In conclusion, we can say that poultry is housed for comfort and protection to achieve efficient production and convenience to the poultry farmer.

References

Agricultural Statistics at a Glance 2021 (1st ed., p. 431). Government of India Ministry of Agriculture & Farmers Welfare Department of Agriculture & Farmers Welfare Directorate of Economics & Statistics. www.agricoop.nic. in&https://desagri.gov.in

Gerzilov, V., Datkova, V., Mihaylova, S., & Bozakova, N. (2012). Effect of poultry housing systems on egg production. *Bulgarian Journal of Agricultural Science*, *18*(6), 953-957.

Oloyo, A., & Ojerinde, A. (2019). *Poultry housing and management*. IntechOpen.

Sastry, NSR., & Thomas, C.K(2023)., Livestock production management, Kalyani Publishers

Singh R.A., (2018). Poultry production., Kalyani Publishers

Yadav, S., & Choudhary, O. Poultry Housing System and Management. *of the Book: Advancement and Innovations in Agriculture*, 207.

CHAPTER-3

DIGESTION AND DIGESTIVE SYSTEM OF POULTRY

[1*]Salur Srihitha, [1]Sikhamoni Haloi and [2]Omprakash Reddy

[1] Assistant Professor, Department of Animal Nutrition, Institute of Veterinary Science & Animal Husbandry, SOA DU, Bhubaneswar, Odisha
[2] MVSc Scholar, Department of Animal Nutrition, CVAS, Mannuthy, Kerala
*Corresponding author- patnaiksrihithaa@gmail.com

Abstract

The poultry digestive system, including the alimentary canal and accessory glands, is highly specialized for efficient nutrient absorption. It begins with the oropharynx and esophagus, leading to the crop for food storage and moistening. The stomach includes the proventriculus for enzyme secretion and the gizzard for grinding food. The small intestine (duodenum, jejunum, ileum) is the primary site for nutrient absorption, while the large intestine reabsorbs water and ferments materials. Intestinal microflora, crucial for digestion and health, varies across different sections. Accessory glands like the liver, pancreas, and salivary glands produce essential digestive enzymes and bile. Efficient nutrient absorption, driven by specialized enzymes and transport mechanisms, ensures the energy needs of poultry are met, supporting their active lifestyles and growth.

Keywords: Poultry, alimentary canal, accessory glands, digestion

Introduction

Understanding the digestive system of poultry is fundamental for optimizing their health, growth, and productivity. Poultry have a unique and highly specialized digestive system designed to efficiently process a variety of feedstuffs. This system, comprising both the alimentary canal and accessory glands, works in concert to ensure rapid digestion and nutrient absorption necessary for the high metabolic demands of birds. This chapter provides a comprehensive overview of the structure and function of each component of the digestive system, from the initial ingestion of food in the oropharynx to the final stages of digestion and nutrient absorption in the intestines. Additionally, the role of intestinal microflora and accessory glands such as the liver, pancreas, and salivary glands in supporting digestive processes will be explored. By delving into the intricate details of poultry digestion, this chapter aims to enhance the understanding of avian physiology and inform better management practices for poultry health and nutrition.

Digestive System:

1. Alimentary canal
2. Accessory glands

1. Alimentary Canal

a. Oropharynx (Mouth and Pharynx)

The oropharynx, extending from the beak to the esophagus, includes the palate, mandible, tongue, and laryngeal mound. The palate has a long median cleft (choana) that connects to the nasal cavity and an infundibular cleft for the auditory tubes. The tongue, controlled by highly developed hyoid muscles, helps move food into the esophagus and fits into the contours of the beak. The oropharynx also houses up to 300 taste buds in chickens and well-developed salivary glands, which, despite having little amylase, provide ample saliva to lubricate and moisten food.

b. Esophagus

The esophagus, with its large size, connects the mouth region to the crop and facilitates the passage of food to the proventriculus. The esophageal wall is composed of four tissue layers, with the mucous membrane being the deepest. This mucous membrane plays a crucial role in defending against microbial invasion and secretes mucus to lubricate food transit through the alimentary canal.

c. Crop:

- The crop serves as a **transient food storage organ** in birds, aiding in biological efficiency by allowing adaptation to varying feeding conditions and maximizing nutrient intake.
- Unlike the limited storage capacity of the proventriculus and gizzard (5-10 g), the crop can manage larger quantities of ingested food, enabling continuous digestion and nutrient absorption.
- The crop gradually moisturizes food to about 50% moisture within 60 minutes, which may activate exogenous enzymes and aid in initial digestion before the food moves further along the digestive tract.

- The pH of the crop varies based on retention time and fermentation, affecting enzymatic activity and overall digestive efficiency, which is essential for nutrient absorption.
- Pigeons produce crop milk from surface cells of the crop to nourish their young during the brooding phase, showcasing a unique adaptation.

d. Stomach (Proventriculus and ventriculus)

- The avian stomach comprises two main parts: the proventriculus and the gizzard. The proventriculus is the glandular or **"true stomach,"** where hydrochloric acid and pepsinogen are secreted to begin digestion. The gizzard, or muscular stomach, is responsible for **grinding feed materials**, aided by its strong muscles and a **koilin layer** that acts like sandpaper.
- Gizzard pH varies with diet and retention time, generally between 3 and 4 for broiler chickens, and can be lowered by adding structural components to the diet, improving digestive function.
- The gizzard retains larger, tougher particles for grinding, allowing smaller particles to pass quickly, ensuring efficient breakdown of coarse feed materials.

e. Small intestine:

- The small intestine in birds consists of the duodenum, jejunum, and ileum. It begins at the end of gizzard and ends at the junction with the caeca and colon, playing a crucial role in **food digestion and nutrient absorption**..
- **The duodenum**, about 20 cm long, forms a loop and contains abundant lymphoid tissue. It receives pancreatic and bile ducts and mixes acidic contents from the gizzard with bile and pancreatic juices, quickly raising the pH for digestion.
- **The jejunum and ileum**, together about 120 cm long, begin at the duodenal end and terminate at the ileo-caecal-colic junction. The jejunum is the main site for digestion and absorption of major nutrients, while the ileum primarily absorbs water and minerals and continues digestion.
- **Villi** in the small intestine increase surface area for nutrient absorption and contain **goblet cells** that secrete mucus, capillaries, nerves, muscle fibers, and lymph vessels.
- The small intestine's functionality and structure, including villus height and intestinal weight, can be influenced by diet and other factors, impacting digestive capacity and nutrient absorption.

f. Large intestine:

- The large intestine in birds consists of the **rectum (colon) and paired ceca, ending at the cloaca**. The bursa of Fabricius, found above the cloaca in young birds, disappears by the age of one year is responsible for immunological activity.

- The ceca, 16-18 cm long in adults, are located at the junction of the small and large intestines. It reabsorb water and ferment coarse materials, producing fatty acids and B vitamins, and aiding in cellulose digestion and handles electrolyte and water absorption, with significant retention time due to infrequent emptying.
- Antiperistaltic movements transport material into the ceca for further processing.
- The rectum (colon) connects the ileum to the cloaca, which is divided into three compartments: **Coprodeum, Urodeum, and Proctodeum.**

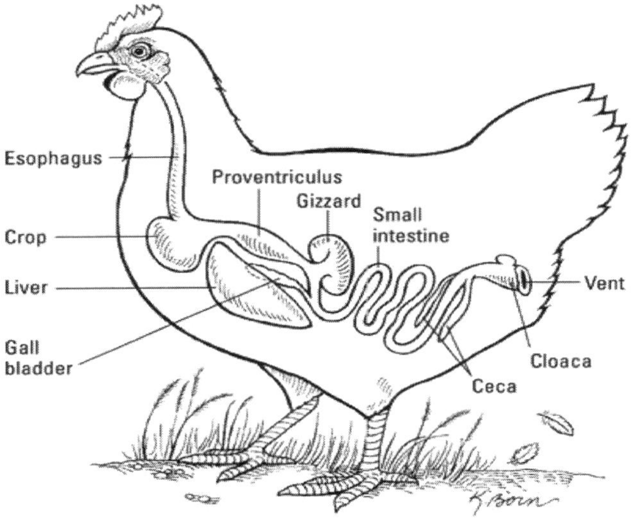

Fig 1. The digestive system of a chicken (Source-Chicken health for dummies)

Intestinal microflora

Microflora refers to beneficial organisms, such as bacteria and yeast, that inhabit the intestines and aid digestion. Chicks have nearly sterile digestive tracts at hatching and naturally acquire beneficial microorganisms from their mother's excrement. Chicks raised via artificial incubation can be given probiotics to supply these beneficial bacteria, helping them fend off harmful germs like salmonella.

Among the three regions of the gastrointestinal tract (duodenum, jejunum-ileum, and cecum), the duodenum has the lowest population of bacteria, while the cecum has the highest. The ceca are populated mainly by Clostridiaceae (65%), with other bacteria including Fusobacterium (14%), Lactobacillus (8%), and Bacteroides (5%). Factors such as dietary protein source, housing conditions, and age affect the bacterial populations in poultry, with antibiotics, enzymes, prebiotics, and probiotics also influencing the dynamics of intestinal flora.

Microflora: Lactobacillus, Enterobacteriaceae, Enterococci, Clostridiaceae, Fusobacterium, Bacteroide etc.

Cloaca and vent

- The cloaca serves as the passage for fecal, urinary, and reproductive materials, with the bursa of Fabricius playing an immunological role.The external opening of the cloaca, known as the vent, varies in size depending on the bird's productivity.

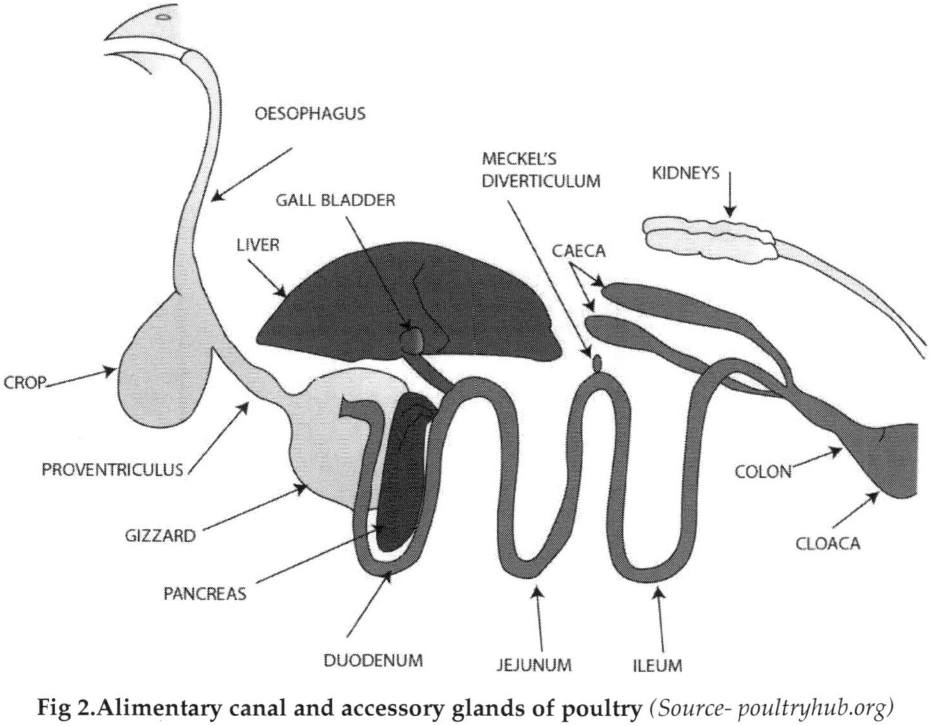

Fig 2.Alimentary canal and accessory glands of poultry *(Source- poultryhub.org)*

Accessory Glands of Poultry

Includes liver, pancreas and salivary glands

a. Liver

The liver, a dark reddish-brown organ, is situated in the right portion of the abdominal cavity, beneath the diaphragm and above the stomach and intestines. As the largest gland in the body, it receives venous blood through the portal vein and arterial blood through the hepatic artery. The liver functions as both an endocrine and exocrine gland.

Functions of the Liver:

- **Bile Production:** The liver produces bile, a substance essential for the digestion and emulsification of fats. Bile is stored in the gall bladder and released into the duodenum as needed for digestion.
- **Metabolism:** It plays a vital role in metabolizing carbohydrates, proteins, and lipids. This includes the storage of glycogen and the conversion of amino acids and fats.
- **Plasma Protein and Fibrinogen Synthesis:** The liver synthesizes essential plasma proteins and fibrinogen, which are crucial for blood clotting.
- **Storage:** It stores glycogen, fats, and fat-soluble vitamins, including vitamin A, which are necessary for various bodily functions.
- **Detoxification:** The liver detoxifies numerous chemicals, removing their toxic effects from the body.

b. Pancreas:

The pancreas produces pancreatic juice with a pH of 6.9, which is released at the distal end of the duodenum loop.

Composition of Pancreatic Juice:

- **Proteolytic Enzymes:** These enzymes break down protein molecules into simpler units and include:
  - » Trypsinogen
  - » Chymotrypsinogen A
  - » Chymotrypsinogen B
  - » Procarboxypeptidase A
  - » Procarboxypeptidase B
- **Lipolytic Enzymes:** These enzymes break down lipid molecules and include:
  - » Phospholipase
  - » Pancreatic lipase
  - » Cholesterol esterase (responsible for the esterification of cholesterol)
- **Carbohydrate-Splitting Enzymes:** These enzymes convert carbohydrates into simpler units and include:
  - » Pancreatic amylase (acts on starch)
  - » Invertase (acts on sucrose)
- **Nucleolytic Enzymes:** These enzymes are also present in pancreatic juice.
- **Cations and Anions:** These act as buffers and osmotic regulators in the pancreatic juice.

c. Salivary glands:

Salivary glands secrete saliva in the mouth, which moistens food, making it easier to swallow and aiding in the initial stages of digestion.

Location of various salivary glands

- Maxillary Glands: Located in the roof of the mouth.
- Palatine Glands: Found on either side of the nasal opening in the roof of the mouth.
- Anterior Sub-Mandibular Glands: Located at the angle formed by the union of the upper and lower beaks or mandibles.
- Posterior Sub-Mandibular Glands: Found in the floor of the mouth.
- Lingual Glands: Embedded in the tongue.
- Crico-Arytenoid Glands: Surround the glottis.
- Apheno-Pterygoid Glands: Situated in the roof of the pharynx, on each side of the common opening for the Eustachian tubes.
- Small Glands: Located in the angle of the mouth.

These glands are covered by a thick layer of stratified squamous epithelium, aiding in the smooth passage of food through the alimentary canal.

Digestion in Poultry

1. Carbohydrate digestion and absorption in birds
2. Protein digestion and absorption in birds
3. Protein digestion and absorption in birds

1. Carbohydrate digestion and absorption in birds

Digestion in the Mouth and Stomach

Carbohydrate digestion in birds begins in the mouth where saliva, containing ptyalin (salivary amylase) and maltase, initiates the breakdown of boiled starch to maltose. In the stomach, no carbohydrate-splitting enzymes are present in the gastric juice, but hydrochloric acid can partially hydrolyze sucrose into glucose and fructose.

Digestion in the Intestine:

In the small intestine, bile does not contribute to carbohydrate digestion, but pancreatic juice releases pancreatic amylase and maltase. Pancreatic amylase acts on both boiled and unboiled starches, breaking them down into maltose. Maltase then further hydrolyzes maltose into glucose. Succus entericus provides additional enzymes such as sucrase, lactase, isomaltase, alpha-limited dextrinase, and intestinal amylase, which digest sucrose, lactose, isomaltose, alpha-limited dextrins, and residual starches, respectively.

$$\text{Starch (polysaccharide)} \xrightarrow{\text{Amylase}} \text{Maltose (disaccharide)}$$

$$\text{Maltose} \xrightarrow{\text{Maltase}} \text{Glucose}$$

Absorption: Monosaccharides are absorbed through passive diffusion (fructose, mannose) or active transport requiring energy and Na+ (glucose, galactose). Among hexoses, galactose is absorbed most rapidly, followed by glucose and fructose. This efficient absorption supports the rapid passage of feed through the avian digestive tract.

2. Protein digestion and absorption in bird

Digestion in the Stomach

- **Proteolytic Enzymes**: The stomach contains three proteolytic enzymes: pepsin, gelatinase, and chymosin.
- Pepsin digests proteins such as albumins and globulins into peptones.
- Proteins → Acid metaprotein → Primary proteose → Secondary proteose → Peptone **(via Pepsin).** Pepsinogen is the precursor of pepsin.
- Nucleoprotein → Nuclein**(via Pepsin)**
- Mucin → Glucosamine + Peptone **(via Pepsin)**
- Gelatin → Gelatin peptone **(via Gelatinase)**
- Caseinogens → Casein **(via Chymosin)**

Digestion in the Small Intestine

- Bile: Contains no proteolytic enzymes.

Digestion in Pancreatic Juice

The pancreas secretes several proteolytic enzymes:

- Proteins/Peptones → Lower peptides/Amino acids **(via Trypsin)**
- Caseinogen → Casein → Polypeptides **(via Chymotrypsin)**
- Polypeptides → Amino acids **(via Aminopeptidase)**
- Tripeptides → Amino acids **(via Tripeptidase)**
- Dipeptides → Amino acids **(via Dipeptidase)**
- Polypeptides → Amino acids **(via Carboxypeptidase)**
- Nucleic acid → Nucleotide **(via Ribonuclease)**
- Elastin proteins → Peptone **(via Elastase)**

- Collagen protein → Peptone **(via Collagenase)**

Digestion in Intestinal Juices

Intestinal juices contain several proteolytic enzymes:

- Polypeptides/Lower peptides → Amino acids **(via Erepsin)**
- Nucleic acid → Nucleotides **(via Polynucleotidase)**
- Nucleosides → Purines + Pyrimidines bases + Pentose phosphate **(via Nucleosidase)**
- Nucleotides → Purines + Pyrimidines bases + Nucleosides **(via Nucleotidase)**

Absorption: Amino acids are absorbed into the bloodstream via the portal circulation, with no peptides entering the portal blood. Absorption rates are highest in the proximal two-thirds of the small intestine, using an active transport mechanism involving sodium. Tripeptides are absorbed faster than dipeptides, which, in turn, are faster than free amino acids, following hydrolysis by peptidases. There is competition among amino acids for absorption,categorized into groups such as acidic, basic, neutral, and imino acids, with natural L-forms absorbed more efficiently than D-forms. Vitamin B6 enhances amino acid transport significantly.

3. Fat digestion and absorption in birds

Digestion in the Stomach

- **Gastric Lipase**: Secreted from the stomach and acts in an acidic pH (4-5). It converts neutral fats into fatty acids and glycerol.

Hydrolysis Steps:

- Step 1: Fat → Fatty acids + Diglycerides
- Step 2: Diglycerides → Fatty acids + Monoglycerides
- Step 3: Monoglycerides → Fatty acids + Glycerol

Overall Equation: 1 Molecule of fat → 3 Moles of fatty acids + 1 Mole of glycerol

Digestion of Fat in the Small Intestine

- Bile, Pancreatic Juice, and Intestinal Juice: Fatty foods are mixed with these substances in the small intestine.

Functions of Bile

- Emulsification: Bile salts (sodium glycocholate and sodium taurocholate) emulsify fat, making fat molecules soluble in water.
- Fat (insoluble) + Bile salts → Emulsified fat (soluble)

Digestion by Pancreatic and Intestinal Juices

- The emulsified fat is digested by enzymes in the pancreatic and intestinal juices to complete the digestion process.

Absorption: Bile salts and phospholipids aid in the formation of mixed micelles, which enhance the absorption of fatty acids, cholesterol particles, and monoglycerides. These micelles are absorbed by the brush border of the intestinal villi, greatly increasing the surface area for absorption. Most of the absorbed components, except bile salts, enter the mucosa by the mid jejunum, while bile salts undergo enterohepatic circulation, being reabsorbed in the ileum. Birds lack a typical lymphatic system, so lipid absorption occurs directly via the portal vein into circulation. This process ensures efficient absorption of essential nutrients and plays a crucial role in the metabolism and energy balance of avian species.

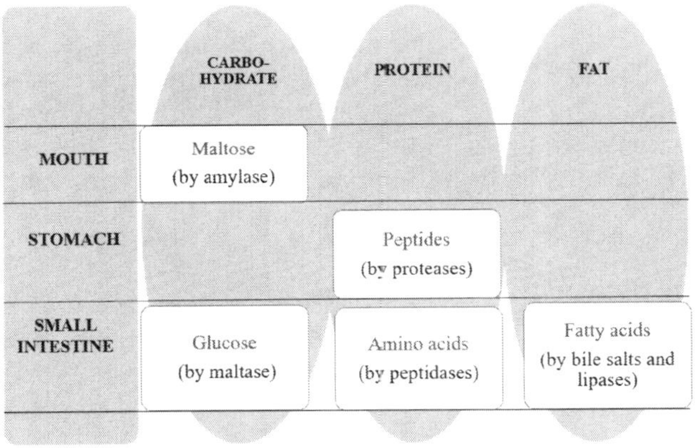

Fig 3. Digestion in poultry

References

Adil, S. and Magray, S.N. 2012. Impact and manipulation of gut microflora in poutry: A review. *J. Anim. Vet. Adv.* 11:873-877.

Albazaz1, R.I. and Büyükünalbal, E.B.2014. Microflora of Digestive Tract in Poultry. *KSU J. Nat. Sci.* 17(1).

Birger Svihus. 2014. Function of the digestive system. *J. Appl. Poult. Res.* 23 :306–314

Dibner, J. J. and Richard, J. D. 2004. The Digestive System: Challenges and Opportunities. *J. Appl. Poult. Res.* 13:86–93.

Neisham, M.C., Austic, R.E. and Card, L.E. 1979. Poultry Production (12th ed.) Lea and Febiger, Philadelphia, USA.

Reece and William. O. 2015. Dukes' Physiology of domestic Animals (13th ed.). Wiley-Blackwell.

CHAPTER-4

FEEDING OF POULTRY

[1*]Sikhamoni Haloi, [1]Salur Srihitha and [2]Nayanmoni Konwar

[1] Assistant Professor, Department of Animal Nutrition, Institute of Veterinary Science and Animal Husbandry, SOADU, Bhubaneswar, Odisha, India

[2] Assistant Professor, Department of Veterinary Microbiology, Institute of Veterinary Science & Animal Husbandry, SOADU, Bhubaneswar

Corresponding author: sikhamonihaloi@gmail.com

Abstract

Poultry feeding is a serious activity that requires attention. Providing a balanced and comprehensive diet to various age groups of chickens and other domestic birds requires specialized knowledge and abilities. There are many factors that affect the amount, composition, and need of nutrients in feed ingredients. Different species of poultry at different stages need a specific diet that is balanced in all aspects and contains a variety of nutrients. Here is a quick rundown of how each of these poultry species is fed. This chapter discusses all feeding practices in broiler and layers in brief. Therefore proper feeding management and practices should be employed for profitable and successful poultry farming

Keywords: Additives, Broiler feeding, Layer feeding, Poultry.

Introduction

Poultry industry is one of the most dynamic sectors of the global agrarian economy, with a large demand for its products because of its widespread social acceptance. In poultry enterprise, feed and feeding management is an important factor of production. Feed accounts to the major cost of poultry (meat and eggs) production, ranging from 65 – 80% under the intensive system of production. The proper feeding management can increase productivity of egg and meat and also reduces cost of production. Nevertheless, inadequate nutrition and feeding practices also result in low egg production, small eggs, poor shell quality, stunted growth, excessive fat accumulation, overfeeding, and a high mortality rate in the poultry husbandry.

Restricted feeding is a cost-effective way for birds to produce more during the laying season, such as larger eggs and longer laying period. It is possible to minimize nutrient loss, effectively utilize feed, and decrease feed waste when feed management is done correctly. Thus, it is critical to have a thorough understanding of feed management practices in order to increase net farm income by feeding nutrients more effectively and to reduce production stress. Some feeding management techniques that can be used to increase profitability in the poultry industry include feed restriction, limited feeding, and adoption of an alternative feed ingredient. The purpose of these feeding management practices is to increase profitability by feeding nutrients more efficiently and to save the stress of production.

Conventional Poultry Feeds

- **Maize:** It has little to no fiber and very easily digested. It serves as an energy source and is poor in sulfur-containing amino acids and protein, particularly lysine. Vitamin A and xanthophyll are abundant in the yellow variety. The latter causes some poultry breeds to have yellow skin.
- **Barley:** Due to its high fiber content, barley is not particularly palatable and should not constitute more than 15% of the ration.
- **Oats:** Due to its high fiber level, oats are not very palatable. It should not constitute for more than 20% of the ration. Due to its manganese level, it might aid in the reduction of cannibalism, hock deformities, and feather picking.
- **Wheat:** As an energy source, Wheat can be utilized in place of maize.
- **Wheat bran:** Because of its fiber content, manganese, and phosphorus, it is bulky and somewhat laxative.
- **Pearl millet:** This feed grain is quite useful and has a nutritional value comparable to that of wheat.
- **Rice:** Broken rice grains can be substituted for maize. This can replace up to 50% of the ration and is a great alternative to cereal grains. Its high oil content makes it prone to going rancid when stored in warm environments.
- **Deoiled rice polish:** Rich in protein and ash content, deoiled rice polish has a low energy content due to the removal of fat.
- **Sorghum:** Sorghum and maize both have comparable feeding values. However, it is palatable, has a higher protein level, and might be substituted for maize. Although costlier than other oilcakes, sorghum meal is a good source of some amino acids.
- **Groundnut-cake:** Often used as a source of protein in poultry feeds, it is highly edible. It has almost 40% protein in it.

- **Fish meal:** One of the best poultry feed ingredients for animal protein is fish meal. Protein content ranges from 45 to 55 percent in most Indian fish meal. Its feeding value is decreased if it contains fish scales.
- **Limestone:** It is a calcium containing feedstuff.. Magnesium content should not be more than 5%.
- **Oyster shell:** An excellent alternative to limestone, oyster shell provides more than 38% calcium. It is also palatable.

Different Form of Poultry Feed

Feed can be given as pellets or as mash. The most widely used method is mash feeding. Wastage of feed decreases when given in the form of pellet and crumble. As in the case of mash-type feed, nutrients are dispersed evenly throughout. For chicks and broilers, crumbled feed works best. Higher feed consumption results in improved growth rates and feed efficiency. Poultry mash feed ought to have a gritty texture and not be excessively fine or coarse.

When feed items are provided individually, birds have the opportunity to choose their own feed. The term "free choice" (cafeteria) system describes this kind of feeding. If given opportunity, a hen is thought to be able to balance her ration.

Fig 1. Different forms of poultry feed (Source:https://www.gemco-energy.com/build-poultry-feed-plant/animal-chicken-poultry-feed-pellets-types-production-process.html)

Different Methods of Feeding Poultry

Grain feeding is not a suitable method of feeding birds because it does not supply all the nutrients needed for the birds' growth and reproduction. Likewise, it is not suitable to feed birds with finely powdered meal. Accordingly, unless the right method is used, a well-balanced ration given incorrectly will not give the best benefits.

Ad libitum Feeding: Also known as free choice feeding, it is simple to manage. The bird has unlimited access to food and is free to control how much it eats overall. Mostly used with broiler chickens, when raising the birds to their maximum body weight is the end goal. Such a method may cause layers and broilers to become obese.

Controlled Feeding: Also referred to as forced feeding. Practiced for a little period of time during the first week when the newborn chicks did not learn to eat. If feed and water are withheld for longer than 36 hours, this technique may be used on other animals to avoid dehydration.

Restricted feeding: Restricted feeding programs are advised for some strains of broiler or laying breeders in order to prevent obesity due to overeating and delay sexual maturity. In broiler, chicks, preferably those that are 18 days or older are fed on a restricted diet for 30 to 35 days to prevent rapid growth of the body that could lead to physiological and metabolic illness syndromes such as ascites and cardio-respiratory failure.

The following benefits come with delayed sexual maturity: (a) a reduction in body weight and fat; (b) a decrease in the cost of producing pullets; (c) an improvement in livability during egg production; and (e) a regulation of egg weight to produce larger eggs.

The different methods of restricted feeding include:

1. Limited everyday Feeding Program: The birds receive a daily amount of feed that is calculated out, but it is less than what they would consume if they were fully fed.
2. Skip-a-day Feeding (Limited Access): The birds are fed every other day, and the quantity of feed provided on feeding days needs to be controlled to twice what is provided on the limited every day program.
3. Changing Nutrient Density: One way to limit nutrient intake is to dilute the ration's nutrient density. This can be achieved by restricting the amount of protein consumed or by lowering energy intake; the latter strategy is predicated on the idea that laying birds have different nutritional requirements at various stages of their productive lives.
4. Phase feeding: For limiting energy intake, phase feeding is typically used. A laying hen's energy needs might range from 310 kcal/ per day in extremely cold weather to 265 kcal/ per day in warm weather.

Phase feeding therefore denotes:

(a) Modifications to the laying hen's diet; Feeding Procedures; (b) adapt to the age and stage of production; (c) adapt to the season, temperature, and climatic changes; (d) take into account the variations in body weight and nutrient requirements of various bird strains; and (e) Take into account changes in one or more nutrients when other nutrients are altered for availability and cost benefits.

Phase feeding involves providing a high-protein diet (often 18–19%) from the start of egg production until the peak of production. For the next five or six months, a low level of protein (about 16%) is fed, and then a further lower level (typically 15%) until the end of the laying season. Phase feeding enables longer-lasting higher and lasting peaks in production. It is well known that with each passing phase, there is a greater need for calcium and a decreased need for phosphorus. Laying type birds are maintained until they are 80 weeks old. The production cycle is separated into three phases, as they are commonly referred to:

Phase-I: The birds should reach their optimum egg production of 85% between the ages of 22 and 36 weeks. In addition, there is a 500 g increase in body weight and a 40 to 60 g increase in egg size. Consequently, for optimal egg production and tissue development, the initial stage of reproduction is essential. In order to provide the necessary amount of protein, the energy level of the poultry feed should be adjusted.

Phase-II: This is the period (37–56 week) when hens have reached adult body weight and their egg production has not dropped below 60%. During this phase, larger eggs are produced, and the protein utilization efficiency is roughly 56%.

Phase-III: It is the phase around 57 and 80 weeks of age, when the production starts to decrease but the body and egg weights marginally increase, or until the spent birds are discarded. Egg production is less than 60% at this stage. Particularly during this stage, there is a decrease in the need for proteins.

Feeding management of birds during summer

Heat stress is more common in poultry birds. Birds in tropical climates suffer greatly from high temperatures and humidity, which had an adverse effect on their performances and ultimately farmer's economy. It may increase water intake and decrease feed consumption. Reduced egg production also affects egg size and shell quality. Additionally, there is a decrease in disease resistance, inadequate feed conversion efficiency, and body weight gain.

With every 10^0C increase in ambient temperature, birds' feed consumption decreases by 1.5%. This decrease in intake of protein and other nutrients leads to a decline in growth and productivity. Providing cool water for birds is usually advised. Thus, throughout the heat, it is imperative to make sure there is an ample supply of cool, clean water. Food should not be given to the birds during the hottest part of the day, when the temperature rises above 37^0C. During the hottest times of the day, wet mash made by combining water can be provided in feeders, which will increase feed intake.

To prevent heat stress, the birds should be given probiotics, vitamins A, E, and C, and antistress medications. It is best to encourage the birds to eat at night and in the early morning. It is not advisable to feed birds in the afternoon during the hot season as this will increase the amount of body heat they have to expel, increasing the risk of heat prostration.

For layers, removing feed before the predicted time of peak temperature is a useful management method to lessen heat stress and relieve the bird of an unnecessary heat load. In order to prevent over activity, broilers can benefit from a time of darkness in the late afternoon. The lighting plan for night feeding should be modified to include an intermittent lighting program. The birds should be fed vitamin C at 330 mg/kg of feed or water to lessen heat stress during the summer.

Feeding management during winter

Birds require more energy during the winter months to keep their bodies warm. Birds will take up more protein and other nutrients along with energy when they eat more feed. As an alternative, the ration's protein and other nutritional content can be decreased while maintaining the same amount of energy. The ideal protein content for broilers during the monsoon season is 22%, although the energy requirement stays around 3100 kcal ME/kg. In winter, the layers needed a greater energy supply (3000 kcal ME/kg) for optimal growth.

Feeding of Broiler and Layer Chicken

Due to their varying dietary requirements, broiler and layer chickens require different diets. Feeding broilers and layers at various stages and age groups is covered in this section.

Broiler Chicken

It is possible to raise broiler stocks quickly; at six weeks of age, they can weigh between 1.8 and 2.0 kg with a FCR of 1.8 kg on whole mash with 90% dry matter or 1.65 kg on pellet diet with 96% dry matter. The following are the main goals of broiler feeding:

(a) Good growth

(b) Better feed conversion ratio (FCR)

(c) Good sound skeletal health and survivability

(d) High-quality meat in a shorter amount of time

Feeding Management in broiler chicks: From the day they are born until the day of sale, broiler chicks are fed ad libitum. It is advantageous to feed broilers often (two to three times) in order to minimize feed waste as well as promote feed intake. It is fed as a whole mash or pellet/crumble feeding.

Fig 2: Broiler Chicken (Source-www.freepik.com)

Growth phases of broiler

The fundamental idea behind phasing out is to supply nutrients based on the type and rate of growth. Commercial broilers are typically raised in three stages over a period of six weeks viz. starter (one to ten days), grower (eleven to twenty-one days), and finisher (22 to marketing age). These birds are raised up to eight weeks of age under particular cases in order to generate larger broiler (roaster) birds that are ideal for producing different value-added products. Feeding regimens have mostly focused on the live performance of meat birds, considering factors including feed conversion, live weight or gain, and occasionally livability. The amount of feed consumed varies across the sexes. The consumption of feed increases with age. There are various specifications available for broiler diet preparation, including BIS, NRC, and CLFMA.

Table 1. Recommendations for typical nutrient levels in the diets of broiler starter, grower and finisher chicken as per BIS (2024)

Nutrients	Requirements for broiler feed		
	Starter	Grower	Finisher
Moisture, per cent by mass, Max	11	11	11
Crude Protein, per cent by mass, Max	22.5	21	19.5
Crude fibre per cent by mass, Max	5	5	5
Ether extract, per cent by mass, Min	3	3.5	4
Acid insoluble ash, percent by mass, Max	2.5	2.5	2.5

Salt (as NaCl), percent by mass, Max	0.5	0.5	0.5
Lysine, per cent by mass, Min	1.3	1.2	1
Methionine, per cent by mass, Min	0.5	0.5	0.45
Calcium, per cent by mass, Min	1	1	1
Total phosphorus, percent by mass, Min	0.8	0.75	0.70
Available phosphorus, per cent by mass, Min	0.48	0.46	0.44
Metabolizable energy (Kcal/kg), Min	3000	3050	3100

Table 2. Some recommended formulae for broiler starter, grower and finisher feeds.

Ingredient	Starter	Grower	Finisher
Maize	34	34.5	38.8
Wheat	10	10	10
Rice polish	5	5	5
Soybean meal	23.5	21.2	16
GNC-SE	10	10	10
Sunflower cake	5	5	5
Animal fat	4.9	6.6	7.6
Fish meal	5	5	5
Calcite	1.1	1.2	1.2
DCP	0.6	0.6	0.65
Salt	0.2	0.2	0.2
Lysine	0.16	0.13	0.04
Methionine	0.14	0.15	0.12
Choline chloride	0.1	0.1	0.1
Coccidiostat	0.05	0.05	0.05
Trace minerals and vitamins	0.2	0.2	0.2
Antibiotic supplement	0.1	0.1	0.1

Source: Reddy (2015)

Layer Chicken

Layers feeding has the following main goals: (a) good growth; (b) good FCR; (c) good skeletal health and survivability; (d) reaching sexual maturity at a specific weight and time with the minimum nutrient intake and (e) increasing egg production.

Feeding management of layer: The purpose of layer ration is to maximize egg production. This can pertain to the egg mass, size, or quantity of eggs. Because of their smaller bodies and slower rates of growth, layer birds consume less feed than broilers. Layer feed is used once the chickens have begun to lay eggs. For chickens that are laying table eggs, layer feeds are formulated. Grower feed is lower in protein and calories than layer feed (BIS, 2024). It also has extra calcium than grower feed, which helps the chickens lay eggs with strong shells.

This feed is given starting at around 20 weeks of age, or sooner if the first egg is deposited. Feeding programs that uses a single feed for the duration of the laying periods are easy but costly. As a result, BIS (2024) suggests splitting the phase-feeding program into phases I and II. Feeding layer birds in different phases allows them to modify their feed intake in line with the rate at which eggs are produced

Fig 3. Layer Chicken (Source-www.freepik.com)

Growth stages of layer

Supplying nutrients according to the type and rate of growth is the basic principle of phasing out. As an example, consider (a) Using hens as breeders, for instance, would produce fertile eggs that would hatch into chicks. (b) Layers: hens raised to produce table eggs for human consumption. The body weight and nutritional needs of the two groups are almost similar but certain micronutrients are provided in excess in breeder diets.

The layers are typically raised in three stages viz. Chicks (0–8 weeks), Growers (9–16 weeks), Layer Phase I (17–18 weeks), and Layer Phase II (19-45 weeks). Layers' nutritional needs are determined by their age and physiological state. Chicks need a ration that can supply the nutrients necessary for quick development of feathers and rapid growth. Rations for chicks are comparatively high in protein, energy, and the vitamins and minerals needed for development and growth. The chicks need less energy once they are completely feathered.

The purpose of layer rationing is to maximize egg production; supply the nutrient needed to maintain health and desired body weight. This may be expressed in terms of the quantity, weight, or size of eggs. The percentage of calcium is increased in layer ration for formation egg shells. The nutrients need, particularly trace minerals and vitamins, is somewhat higher for breeders since, in addition to egg production, hatchability and fertility are also significant factors.

Table 3. Recommendations for different stages of layer as per BIS (2024)

Nutrients	Requirements for Layer feed			
	Chick	Grower	Layer phase I	Layer phase II
Moisture, per cent by mass, Max	11	11	11	11
Crude Protein, per cent by mass, Max	20	16	18	16
Crude fibre per cent by mass, Max	7	9	9	10
Ether extract, per cent by mass, Min	2	2	2	2
Acid insoluble ash, percent by mass, Max	4	4	4	4.5
Salt (as NaCl), percent by mass, Max	0.5	0.5	0.5	0.5
Lysine, per cent by mass, Min	1	0.7	0.7	0.65
Methionine, per cent by mass, Min	0.4	0.35	0.35	0.3
Calcium, per cent by mass, Min	1	1	3.75	4
Total phosphorus, percent by mass, Min	0.7	0.65	0.65	0.65
Available phosphorus, per cent by mass, Min	0.45	0.40	0.35	0.33
Metabolizable energy (Kcal/kg), Min	2800	2500	2600	2400

Table 4. Some recommended formulas for practical rations for various stages of layer

Ingredient	Chick	Grower	Layer phase I	Layer phase II
Maize	35	20	24.5	25
Jowar	0	0	22	0
Broken rice	10	0	0	10
Bajra	0	25.7	0	10
Rice polish	15	0	4	10
DORB	8	25	15	15.4
Wheat bran	0	5	0	6
Sunflower cake	0	10	0	10
GNC-SE	7	0	10	0
Soybean meal	17.5	7	11.3	9
Fish meal	5	5	5	5
calcite	1.4	1.6	7.4	9

Ingredient	Chick	Grower	Layer phase I	Layer phase II
DCP	0.3	0	0.15	0
Salt	0.2	0.2	0.2	0.2
Lysine	0.08	0	0	0
Methionine	0.07	0.04	0.05	0
Choline chloride	0.1	0.1	0.1	0.1
Coccidiostat	0.05	0.05	0.05	0.05
Trace minerals and vitamins	0.2	0.2	0.2	0.2
Antibiotic supplement	0.1	0.1	0.1	0.1

Source: Reddy (2015)

Feed supplement and additives in poultry

Supplements are added to animal feed to provide essential nutrients that may be missing or insufficient in the standard ration. While feed ingredients supply energy and protein, they often lack adequate amounts of certain minerals, vitamins, and amino acids.

- **Mineral Supplements**: These are added in synthetic or organic complex forms, like chelated minerals, to improve absorption and correct deficiencies.
- **Vitamin Supplements**: Synthetic forms of vitamins are included to meet the animal's nutritional needs when the feed ingredients do not provide enough.
- **Amino Acid Supplements**: Limiting amino acids, such as DL-methionine and L-lysine, are supplied in synthetic forms to support protein synthesis and growth.

By incorporating these supplements, livestock receive a well-balanced diet, which enhances their health, growth, and productivity.

Feed additives play a significant role in poultry nutrition by enhancing various aspects of feed intake, digestion, and nutrient utilization. Here's a brief explanation of the key types of poultry feed additives

1. **Antibiotics**: These additives are used to prevent or treat infections and diseases in poultry. They also help promote growth by improving gut health. Example-Chlortetracycline, Zinc bacitracin, Virginiamycin etc.
2. **Probiotics**: Live beneficial microorganisms that enhance gut flora, improving digestion and nutrient absorption, thereby contributing to the overall health and productivity of the birds. Example- *Lactobacillus sp*, *Streptococcus sp* etc.

3. **Prebiotics**: Non-digestible ingredients that stimulate the growth and activity of beneficial gut bacteria, promoting a healthy digestive system and improved nutrient utilization. Example- Mannan oligosaccharide, Fructose oligosaccharide etc.
4. **Acidifiers**: Organic or inorganic acids added to the feed to lower the pH in the gastrointestinal tract, improving feed digestion and reducing pathogenic bacterial growth. Example- Formic acid, Propionic acid, Fumaric acid etc
5. **Enzymes**: These are added to the feed to help break down complex nutrients, such as fibers, proteins, or starches, making nutrients more available for absorption. Enzymes- Xylanase, Beta glucanase etc.
6. **Emulsifiers**: Compounds that help improve the digestion and absorption of fats by breaking down fat molecules, leading to better nutrient uptake and energy utilization. Example- Lecithin, Glyceryl monostearate etc.
7. **Toxin binders**: Substances added to the feed to bind and neutralize mycotoxins and other harmful substances, ensuring the feed is safe for consumption and reducing the risk of toxin-related health issues. Example- Hydrated Sodium Calcium Aluminosilicate (HSCAS) and Zeolites.
8. **Antioxidants**: These additives prevent oxidative damage to fats and vitamins in the feed, ensuring the feed remains fresh and nutritious. Example- Butylated hydroxyl anisole (BHA), Butylate hydroxyl toluene (BHT), Ethoxyquin etc.
9. **Herbal growth promoters**: Natural plant extracts used to stimulate growth, improve immunity, and enhance feed conversion efficiency in poultry without the use of synthetic chemicals. Example- Aloevera, Neem, Garlic, Turmeric etc.
10. **Coloring pigments**: Additives used to enhance the color of poultry products like skin or egg yolks, which are often preferred by consumers for aesthetic reasons. Natural pigments like carotenoids are commonly used. Example- Xanthophylls, Arsanilic Acid, Sodium arsanilate etc.

These additives, when used appropriately, can significantly boost the performance and health of poultry while improving the quality of the products.

Use of Alternative Feed Ingredients

In order to make livestock rearing economically viable, it is necessary to investigate the use of alternatives to the costly conventional feed resources, such as maize, corn, and oil seed cakes, due to their rising costs. Various scientific studies have been conducted on a number of alternative feed resources found in the tropics. Alternative feed sources are being used in place of conventional ones in various locations that have not been fully exploited but have been found to be beneficial for animal feed. Some alternative feed resources used by some researcher are given below.

Table 5. Examples of non -conventional energy feed resources

By product	Example
Grains	Wheat bran and middling, sorghum bran, rice bran and husk, haulms as well as brewers' dried grain
Roots and tubers	Cassava peels, yam peels, potato peels and cocoyam peels
Animals	Meat meal, bone meal, and blood meal
Fats and oils	Palm oil slurry, soap stock, and vegetable fats
Miscellaneous	Molasses, syrups, sugar, sugar cane scrapings

Source: Ogunnusi *et al.*, (2023)

Table 6. Examples of non-conventional protein feed resources

Ingredients	Crude protein (%)
Blood meal	77.35
Feather meal	85.00
Animal manure	20.00
Poultry waste	25.00
Maggot meal	54.00
Earthworm	65.00
Glyricidia sepium leaf	32.00

Source: Ogunnusi *et al.*, (2023)

Storage of Poultry Feed

The following steps should be followed for appropriate storage in order to avoid deterioration and waste of feed:

1. The storehouse should be dry and impervious to rodents.
2. It is important to keep wild birds out of the storehouse.
3. Adequate lighting and ventilation are essential for the store house.
4. Disinfectant tubs have to be used at the feed store entry.
5. Maintain a clean and tidy feed store at all times.
6. A sufficient disinfectant must be fumigated into the feed store at the appropriate intervals.
7. The ingredient and feed bags need to be set on wooden or iron slats.
8. In order to avoid contaminating feed, the store house needs to be built apart from the chicken house.

Conclusion

In order to feed balanced rations more effectively and reduce production stress, farmers can increase their net income by having a solid understanding of feeding management techniques. Practicing feeding management strategies like as provision of balanced meal, limiting feeding, adopting alternative feed ingredients, and feed restriction can help poultry famers to ensure profitable and effective poultry farming

References

BIS. [Bureau of Indian Standards] 2024. IS 1374: Poultry Feeds – Specification (Sixth revision).

Ogunnusi, O. J., Toye, C. O., and Akinwemoye, A. A. (2023). Feed management as a paradigm for profitable poultry enterprise. Animal Research International, 20(1), 4684-4693.

Oyedeji, J. O., Orheruata, A. M. and Omatsuli, M. (2007). Effects of feed rationing on the laying performance of 40-weeks in-lay hens. Journal of Food Agriculture and Environment, 5(3/4): 301 – 303.

Poultry Production and Management: Increasing Egg, Meat Output, Addressing Feed Challenges, 2nd Edition (2024) Dr P. P. Singh Dr. Gaurav Jain, Dr. Aslam , Dr. Satendra Kumar (Author) (ISBN 978-81-963487-9-3)

Reddy, D. V 2015. Applied Nutrition Livestock, Poultry, Rabbits and Laboratory Animals. 3rd ed. Oxford and IBH Publishing Co. Pvt. Ltd, New Delhi. pp: 189-225.

CHAPTER-5

RATION FORMULATION FOR POULTRY

[1*]Sikhamoni Haloi and [1]Salur Srihitha

[1] Assistant Professor, Department of Animal Nutrition, Institute of Veterinary Science and Animal Husbandry, SOADU, Bhubaneswar, Odisha, India

*Corresponding author: sikhamonihaloi@gmail.com

Abstract

This chapter provides a comprehensive overview of poultry ration formulation, emphasizing the importance of balancing cost, nutritional quality, and ingredient flexibility. Various formulation methods are discussed, such as Pearson's Square and algebraic method alongside practical steps for creating a balanced ration. The chapter illustrates calculations for crude protein and metabolizable energy using animal protein, vegetable protein and energy sources.

Keywords: Poultry, Poultry Ration, Broiler ration

Introduction

This chapter outlines the principles of poultry feed formulation, focusing on meeting nutritional requirements while balancing cost and ingredient availability. It covers key factors like the nutritional value of ingredients, feed additives, and testing for contaminants. Various methods, such as Pearson's Square and computerized formulation, are used to balance crude protein, energy, and other nutrients. Practical steps ensure an optimal, cost-effective diet tailored to the specific needs of poultry based on age, weight, and breed.

Important points to consider in poultry feed formulation

- Accessibility and cost of feed components
- Flexibility in feed formulation – optimal inclusion levels
- Palatability and acceptability
- Nutritional quality and value of feed ingredients
- Nutritional requirements for poultry based on age, weight, breed, and class
- Use of feed additives – including anti-stress agents, non-vitamin supplements, growth enhancers, coccidiostats, antibiotics, etc.
- Testing for aflatoxins, mycotoxins, and other feed contaminants

Poultry rations can be formulated using the following methods:

1. **Pearson's Square Method** – A simple technique for balancing two ingredients based on their nutrient content.
2. **Algebraic Method** – Involves solving equations to calculate the appropriate proportions of multiple ingredients.
3. **Linear Programming** – A mathematical approach used to optimize feed formulation while minimizing cost and meeting all nutrient requirements.
4. **Trial and Error Method** – A practical approach, adjusting ingredients based on observation and performance.
5. **Computerized Feed Formulation** – Uses specialized software to calculate nutrient requirements and formulate balanced rations efficiently.

These methods vary in complexity and accuracy, depending on the specific needs and resources.

Steps in feed formulation

Let's say we need to prepare a 100 kg ration for a broiler grower, following the nutrient requirements outlined by BIS 2024: Crude Protein (CP) = 21% and Metabolizable Energy (ME) = 3050 kcal/kg of feed. The key steps in practical diet formulation through calculations are as follows.

Step I: Minor ingredients are either fixed or a slack space (5 kg) can be left for later inclusion. Trace minerals, vitamins, and feed additives can be predetermined because their contribution from major feed ingredients is minimal. Slack space may also be reserved for the later addition of salt, calcium, phosphorus sources, supplemental amino acids, and fat to balance the diet.

Step II: The levels of animal protein sources can be fixed. These can be added to fixed level because of the cost. These ingredients provide the limiting amino acids like lysine, cystine and methionine. Fish meal = 10.00 kg

Step III: The level of cereal by-products like rice polish, rice bran, wheat bran if to be added may be fixed. Alternatively, the cereal by-products and cereals can be fixed in certain ratio and can be added. Rice polish = 10.00 kg

Table 1. Nutrient composition of the ingredients used

Ingredients	CP (%)	ME (Kcal/kg)
Maize	9	3300
Soybean meal	45	2400
Rice polish	13	2937
Fish meal	45	1834

Now calculate the CP provided by above quantity

Slack space	5	———	0
Fish meal	10	45×10/100	4.5
Rice polish	10	13×10/100	1.3
Total	25		5.8

Step IV: Vegetable protein and energy sources are added to meet the required protein level. So far, 5 kg of ingredients have been included. Till now 25.00 kg of ingredients were added (including the slack space) and these contributed 5.8 percent protein. Still (100-25=75) 75.00 kg of ingredients are to be added and 15.20 % (21-5.8) protein is to be provided because total CP in the diet should be 21.05 %.

- So far, 25.00 kg of ingredients have been added, including slack space, contributing 5.8% protein. There are still 75.00 kg of ingredients left to add (100 - 25 = 75), which must provide an additional 15.20% protein (21 - 5.8) in order to reach the target of 21% crude protein (CP) in the final diet.
- Soybean meal as vegetable protein source, along with maize as an energy source, are considered. The required protein content can be calculated using Pearson's formula or algebraic methods.

Pearson's formula

Total quantity of ingredients = 75 kg

Protein = 15.20 kg

Protein as percent in = 15.20 X 100/75 = 20.27 %

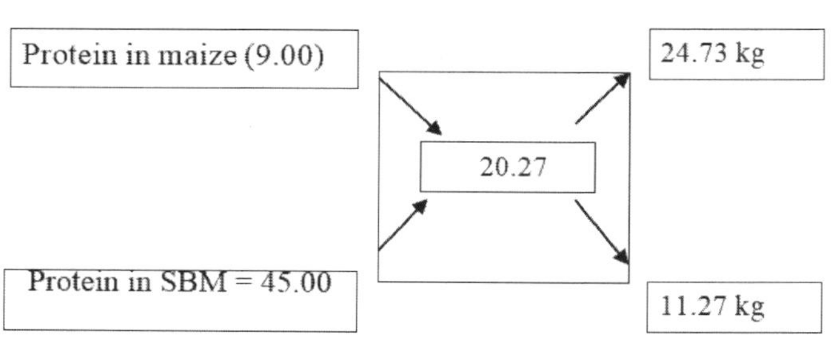

36.00 kg

- The desired protein percentage is placed in the center of a square. The protein percentages of the vegetable protein sources (Soybean meal) and maize are entered at the corners. The difference between these protein percentages determines the ratio of the ingredients to be mixed.
- Proportion of maize in 75 kg= 24.73/36 × 75 =51.52 kg
- Proportion of SBM in 75 kg= 11.27/36 × 75= 23.48 kg

Or

Algebraic Equation:

Total of ingredients = 75 kg

Protein = 15.20 kg

Let 'x' represent the protein source and 'y' represent maize.

The protein content of protein source SBM is 45.0 % and Maize 9 %.

x + y = 75.00 — I

0.45 x + 0.09 y = 15.20 — II

0.09x + 0.09 y = 6.75 — III (Equation I X 0.09)

0.36x = 8.45

x = 8.45/0.36

x = 23.48kg

y = 75.00 – 23.47 = 51.52 kg.

Ingredient	Parts	CP%	ME(Kcal/kg)
Maize	51.52	4.64	1700.16
SBM	23.48	10.57	563.52
Total	75.00	15.21	2263.68

Table 2. Final Formula for broiler ration:

Ingredient	Parts	CP%	ME(Kcal/kg)
Maize	51.52	4.64	1700.16
SBM	23.48	10.57	563.52
Fish meal	10	4.5	183.40
Rice polish	10	1.3	293.70
Slack space (vitamins, minerals, feed additives and salt)	5	-----	-------
Total	100	21.01	2740.78
Shortfall			309.22

Step V: The metabolizable energy (ME) content of the diet must be balanced. The shortfall of 309.22 kcal (3050.00 – 2740.78) can be compensated by adding animal fats or vegetable oils. However, the inclusion of fat depends on economic feasibility and may not always be practical.

Step VI: The phosphorus content of the diet is calculated in terms of available phosphorus. Phosphorus from animal and inorganic sources is considered fully available, while phosphorus from plant sources is assumed to be 30% available.

Step VII: The calcium requirement can be fulfilled by adding limestone and bone meal.

Step VIII: Limiting amino acids in synthetic form can be supplemented to meet the nutritional requirements. The decision to use synthetic amino acids is based on cost considerations. Alternatively, the need for these amino acids can be met by increasing the inclusion of animal protein sources.

Step IX: A final check is done to ensure that the total weight of the ingredients equals 100 kg, and that all required nutrients are properly balanced. If the total does not reach 100, additional cereal or cereal by-products can be added to complete the diet.

Conclusion

In conclusion, effective poultry feed formulation is critical for ensuring optimal growth, health, and production. By considering factors like ingredient cost, nutritional content, and feed additives, along with using various formulation methods, a balanced and cost-efficient diet can be achieved. Tailoring feed to the specific needs of poultry helps maximize performance, while careful attention to energy, protein, and other nutrient requirements ensures overall flock health and productivity.

References

Reddy, D. V 2015. Applied Nutrition Livestock, Poultry, Rabbits and Laboratory Animals. 3rd ed. Oxford and IBH Publishing Co. Pvt. Ltd, New Delhi. pp: 189-225.

Ranjhan, S. K. 2004. Animal nutrition and feeding practices. Sixth edn., Vikas Publishing House Pvt. Ltd., New Delhi, India

CHAPTER 6

POULTRY REPRODUCTION AND BREEDING

[1*]Mekapothula Hima Bindu, [2]Chandaka Deepthi and [3]Salur Srihitha

[1] Assistant Professor, Department of Animal Genetics and Breeding, Institute of Veterinary Science & Animal Husbandry, SOA DU, Bhubaneswar, Odisha
[2] Assistant Professor, Department of Livestock Farm Complex, School of Veterinary & Animal Sciences, Centurion University of Technology & Management, Paralakhemundi, Odisha
[3] Assistant Professor, Department of Animal Nutrition, Institute of Veterinary Science & Animal Husbandry, SOA DU, Bhubaneswar, Odisha
**Corresponding author- himabindu9617@gmail.com*

Abstract

Reproduction in poultry involves distinct anatomical and physiological adaptations compared to mammals. Male chickens have internally located testes and a simple duct system, including the vas deferens and female chickens have a specialized reproductive system for egg production. The egg formation process, from ovulation to laying, takes about 23-26 hours, with yolk development, albumen addition, shell membrane creation, and shell formation occurring sequentially. Mating involves semen deposition in the cloaca, with sperm stored in sperm-storage glands for extended periods, ensuring fertilization in the infundibulum. Poultry breeding systems aim to enhance homozygosity or heterozygosity, using methods like random mating, inbreeding, line breeding, outbreeding, and various crossbreeding techniques. Selection is critical for genetic improvement, involving individual, family, and pedigree selection. Specific methods include tandem, independent culling, and selection index, with an ideal breeding program combining these methods. Selection of birds for breeding prioritizes body conformation, performance, and genetic traits. Mating methods include individual mating, pen mating, flock mating, and artificial insemination. These systems and methods cater to different breeding goals, optimizing genetic diversity and enhancing desirable traits in poultry populations.

Keywords: Avian, Selection, Egg, Recurrent, Yolk, Breeding

Introduction

The avian reproductive system is characterized by heterosexual reproduction, necessitating both a male and a female to each contribute half of the genetic material for the offspring. It involves a unique set of anatomical and physiological adaptations that differ from those of mammals. The male provides sperm through the testes, transported in semen, while the female offers the ovum within the egg yolk, produced by the ovary. Here's an overview of the reproductive process in chicken.

Reproduction in poultry

Poultry reproduction involves a unique set of anatomical and physiological adaptations that differ from those of mammals. Egg formation is complex, with genetic and management factors influencing quality. Understanding this system aids farmers in maximizing egg production and quality.

Male Reproductive System

The male reproductive organs in domestic fowl include **two testes** and deferent ducts leading from the testes to the cloaca (Figure 1). Unlike many other animals, fowls lack a penis. The bean-shaped testes are situated against the backbone at the front of the kidneys, varying in size with sexual activity. Typically, the left testis is larger than the right. Each testis has a small, flattened area thought to be analogous to the **epididymis** in mammals.

The Vas deference duct originates here and transports sperm from the testes to the cloaca, from where it enters the female's oviduct during mating. The **vas deferens** enlarges into a bulbous region before entering the **cloaca**, which serves as a storage site for semen. The paired vasa deferentia terminate in a double, teat-like structure known as **papillae**.

In the testes, **seminiferous tubules** conduct meiosis and transformation processes to produce sperm, which carry half the chromosomes required for an embryo. Each cubic millimeter of semen contains 3-5 million sperm, visible under a microscope with long, pointed heads and tails. The testes also produce androgens, influencing secondary sex characteristics like comb growth, male behavior, and mating.

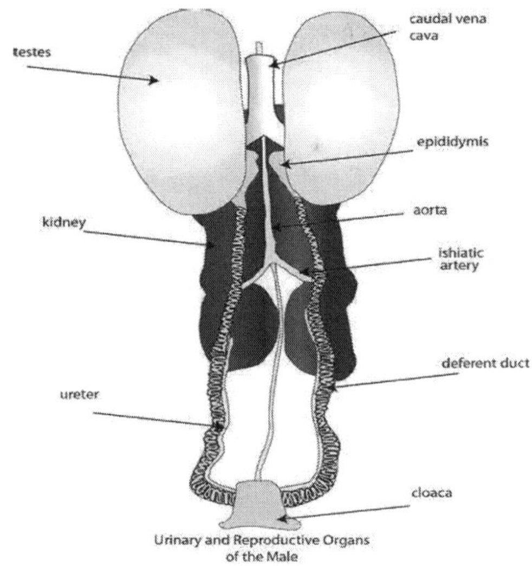

Fig 1: Male Reproductive System (Image source: Poultry Hub)

Female Reproductive System

The female reproductive system in domestic fowl consists of the ovary and the accompanying oviduct (Figure 2). While female chicken embryos have two sets of reproductive organs, only the left side matures to produce eggs. The functional ovary is located in front of the kidneys in the abdominal cavity, firmly attached to the cavity wall and richly supplied with blood vessels to transport nutrients to the developing yolk.

- **Ovary**: The ovary releases eggs, which are then received by the infundibulum, the upper part of the oviduct. The ovary comprises numerous yellowish follicles, each containing an ovum or yolk. Though many follicles exist, only a small proportion mature to produce eggs. The follicles vary in size, from very small to those reaching up to 40 mm in diameter, containing a fully matured yolk ready for release into the oviduct.
- **Oviduct:** The oviduct's function is to add albumen, shell membranes, and the shell around the yolk. This long tube, rich in blood vessels and glands, grows to 70-80 cm in a laying hen and varies in diameter depending on the section's function.

The oviduct has five distinct sections, each with specific roles: (Table 1)

1. **Infundibulum:** Adjacent to the ovary, it collects the yolk and directs it into the oviduct. Fertilization occurs here.
2. **Magnum (Ampulla):** Approximately 40 cm long, it secretes over 40% of the albumen (egg white).

3. **Isthmus:** About 12 cm long, it secretes some albumen and the shell membranes (both inner and outer shell membranes).
4. **Uterus (Shell Gland):** Around 12 cm in length, it secretes about 40% of the albumen and the egg shell.
5. **Vagina (Cloaca):** Connects the uterus to the cloaca. It serves as a passageway for the egg and also receives semen during mating.

Table 1: Parts of Oviduct

Oviduct Parts	Length	Time Spent	Functions
Infundibulum	2 Inches	15 min	Yolk and egg fertilization
Magnum	13 Inches	3 hrs	Thick white albumin
Isthmus	4 Inches	1.25 hrs	Albumin shell membranes Shape of egg was determined
Uterus	4.2 Inches	20.75 hrs	Egg shell Pigment for egg colour
Cloaca	4 Inches	-	Egg passes

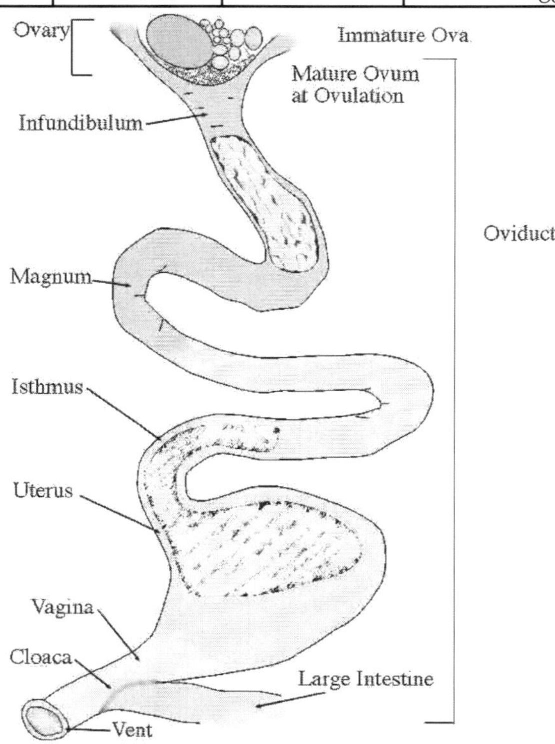

Fig 2: Female Reproductive Tract (Image source: pininterest.co.uk)

Formation of the hen's egg:

The formation of a hen's egg is a complex and coordinated process involving several key components and stages. Here's an overview of how each part of the egg is formed:

Key Components of the Hen's Egg

- **Yolk (Ovum):** Produced by the ovary.
- **Albumen (Egg White):** Primarily produced in the magnum.
- **Shell Membranes:** Formed in the isthmus.
- **Shell:** Created in the uterus or shell gland.

a. Yolk Formation

- **Ovary Structure:** The ovary is connected to the abdominal cavity wall by the meso-ovarian ligament and contains 2,000 to 12,000 small ova (immature eggs) within follicles. Only 200 to 350 of these ova mature under typical commercial conditions.
- **Yolk Development:** The yolk is formed in concentric layers of material, with its color coming from xanthophylls (yellow/orange/red pigments). It serves as the main nutrient source for the developing embryo. The yolk contains approximately:
  - Water: 48.0%
  - Protein: 17.5%
  - Fat: 32.5%
  - Carbohydrate: 1.0%
  - Other Compounds: 1.0%
- **Follicle Formation:** The yolk is housed in a follicle on the ovary, which is well-supplied with blood vessels to transport yolk constituents from the liver.
- **Hormonal Regulation:** Yolk development is stimulated by follicle-stimulating hormone (FSH) from the pituitary gland. This hormone prompts the liver to produce yolk materials, which are then transported to the ovary. Ovarian hormones like estrogen, progesterone, and testosterone, also triggered by FSH, contribute to the development of the oviduct.

b. Ovulation

- **Process:** Ovulation is the release of the mature yolk from the follicle into the oviduct. It is triggered by luteinizing hormone (LH) from the pituitary gland. This hormone causes the follicle to rupture and release the yolk.
- **Timing:** The yolk is released and enters the oviduct, where fertilization (if it occurs) and further egg development take place.

c. Albumen Formation

- **Location:** The albumen (egg white) is mainly formed in the magnum, the longest part of the oviduct.
- **Process:** As the yolk travels through the oviduct, layers of albumen are deposited around it. The albumen provides additional protection and nourishment.

d. Shell Membranes Formation

- **Location:** The shell membranes are formed in the isthmus, which is the segment of the oviduct between the magnum and the uterus.
- **Process:** Two shell membranes, an inner and an outer membrane, are created around the developing egg.

e. Shell Formation

- **Location:** The shell is formed in the uterus, also known as the shell gland.
- **Process:** In the uterus, layers of calcium carbonate are deposited around the egg, forming the hard shell. This process takes about 20 hours and ensures the egg's protection and structure.

Sexual Maturity and Egg Laying

- **Sexual Maturity:** Hens reach sexual maturity and lay their first eggs between 18 and 24 weeks of age. This timing can be influenced by genetics and environmental factors like lighting and feeding.
- **Ovulation Cycle:** Once the hen starts laying eggs, the timing of ovulation is closely linked to the laying pattern. Subsequent ovulations occur approximately 40-60 minutes after the previous egg is laid.
- **Clutches:** A clutch is a group of eggs laid in succession, typically ranging from 3 to 8 eggs. Larger clutches usually indicate better production performance.

Egg Formation Time

- **Duration:** The entire process from ovulation to egg laying takes about 23 to 26 hours. If it exceeds 24 hours, subsequent eggs in the clutch are laid progressively later.
- **Laying Pattern:** Initially, egg laying can be irregular. Over time, as the hen's hormonal systems stabilize, egg production becomes more consistent, peaking 3-5 weeks after the first egg and then gradually declining.

This intricate process ensures the development of the egg's components in a sequential manner, culminating in the formation of a fully-formed egg ready for laying.

Mating and fertilization

- **Mating Process:** During mating, the male deposits semen in the female's cloaca. In chickens, the semen is stored in the sperm-storage glands or crypts on the vaginal side of the uterovaginal junction (Figure 3).
- **Sperm Storage:** Sperm can be stored in these crypts for extended periods.
- **Fertilization:** Fertilization occurs in the infundibulum. Sperm from the storage crypts ascend the oviduct to meet the ovum, assisted by peristaltic contractions.

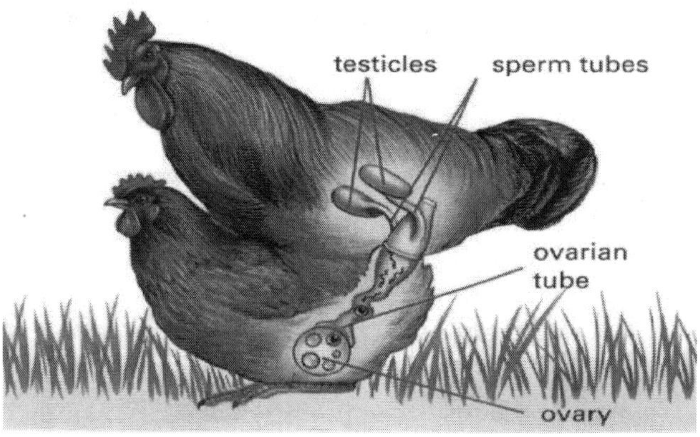

Figure 3: Mating in chicken *(Image source: Bitchin'chickens)*

Mating methods

- **Individual Mating (Stud Mating):** Each female is mated individually with a male, which allows for precise control of parentage but is labor-intensive.
- **Pen Mating:** A male is kept with a group of females, allowing for pedigree maintenance and identifying parentage but may have lower fertility rates.
- **Flock Mating:** Multiple males are mated with a flock of hens, promoting frequent matings and generally higher fertility but lacks pedigree control.
- **Artificial Insemination (AI):** The semen is collected and manually introduced to females. This method ensures optimal fertility and is useful in controlled environments.

These breeding systems and methods provide various approaches to managing poultry genetics, each with its own benefits and challenges tailored to different breeding goals.

Poultry breeding

The suitability of a male or female bird for breeding is largely determined by the quality of their progeny, regardless of the trait in question. For instance, a male whose mother had high egg production, when mated with a high-producing female, typically produces daughters that also lay prolifically. The outcomes of a mating are more influenced by the genetic contributions of the paired birds than by their physical appearance.

Breeding systems

Breeding systems are classified based on their aim to increase homozygosity or heterozygosity. These systems include random mating, inbreeding (which increases homozygosity), and outbreeding (which increases heterozygosity).

A. Random Mating

Random mating involves pairing individuals without any selection criteria, meaning any individual of one sex has an equal chance of mating with any individual of the opposite sex. This method is primarily used to create a control population for comparison and evaluation of other breeding systems.

B. Breeding for Increased Homozygosity

Inbreeding

Inbreeding refers to mating individuals that are more closely related than the average relationship in the population. This can be performed through several distinct methods:

- **Close Inbreeding:** Mating between siblings or between parents and offspring, often used to develop inbred lines. This reduces heterozygosity significantly over generations.
- **Strain Formation:** Establishing a small group within a breed or variety with specific characteristics. For example, the Babcock strain of Single Comb White Leghorns was developed to enhance egg weight.
- **Line Breeding:** Intensive inbreeding within an ancestral line, achieved through repeated back-crossing. This aims to solidify certain traits over several generations.

Mating Based on Phenotypic Resemblance: Pairing individuals with similar traits without considering their genetic relationship. This is useful when the heritability of the traits in question is high.

C. Breeding for Increased Heterozygosity

Outbreeding

Outbreeding involves mating individuals that are less closely related than the average of the population.

This approach includes various methods:

- **Single Two-Way Cross:** Crossing two different populations (inbred lines, strains, or breeds) to produce an F1 generation for commercial purposes.

These hybrids often show hybrid vigor.

C X D

↓

CD

- **Three-Way Cross:** Mating F1 cross-bred females with males from a third line or breed to maximize hybrid vigor.

C ♀♀ X D ♂♂

↓

CD ♀♀ X E ♂♂

↓

CDE

- **Four-Way or Double Cross:** Combining two single crosses (e.g., CD x EF) to produce a new population with potentially enhanced traits.

C X D   E X F

↓   ↓

CD   EF

↓

CDEF

- **Back-Cross:** Mating F1 females with males from one of the parental populations, useful for improving maternal traits.

♀ Parent C X R male parent

↓

F1 ♀ D X R

↓

E

- **Rotational Cross-Breeding:** Alternating males from different populations each generation to maintain genetic diversity.

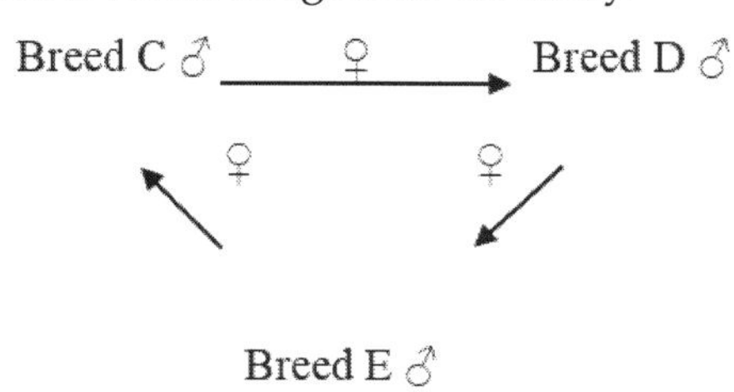

Example of Rotational Cross-Breeding

Let's consider a rotational cross-breeding scheme involving three breeds: C, D, and E.

- **Generation 1:** Breed C x Breed D = F1 Generation
- **Generation 2:** F1 Generation x Breed E = F2 Generation
- **Generation 3:** F2 Generation x Breed C= F3 Generation
- **Generation 4:** F3 Generation x Breed D = F4 Generation

In this example, the rotation plan alternates between breeds C, D, and E, ensuring that each breed contributes to the genetic pool at different points in the cycle. This rotation helps to combine the strengths of each breed and maintain genetic diversity within the population.

Overall, rotational cross-breeding is a strategic approach to poultry breeding that leverages genetic diversity and hybrid vigor while minimizing the risks associated with inbreeding

- **Top Crossing:** Mating inbred males with females from non-inbred populations to combine advantageous traits.

Inbred♂ X Non-inbred

↓

Top cross

- **Grading Up:** Back-crossing to a superior breed over several generations to convert a mixed breed population to a pure breed.
- **Crossing for New Breed Production:** Crossing different breed types to create new breeds that combine desirable traits, followed by inbreeding and selection to stabilize the breed.
- **Diallel Mating:** All possible crosses among multiple strains to identify the best combinations for hybrid vigour and general combining ability.

↘ • ♂ • ♀	• A	• B	•	• D
• A	• A A	• B A	•	• D A
• B	• A B	• BB	•	• D B
• C	• A C	• BC	•	• D C
• D	• A D	• B D	•	• D D

Ideal breeding programme

An ideal breeding programme combines various selection systems. This integrated approach utilizes individual, family, and pedigree selection systems to optimize results. Traits such as egg production, viability, and fertility, which have low heritability, can be improved through family selection. Conversely, traits like egg weight, shell quality, sexual maturity, growth rate, and conformation can be effectively enhanced using individual selection. A skilled breeder combines these methods to achieve the desired characteristics in their birds.

Selection of birds for breeding

Egg Type Lines:

The optimal age for selecting egg-type birds is between 10 and 14 weeks, during which inferior birds are culled. Body conformation and the development of structural body parts are prioritized over body weight in selecting breeders. Birds from each line are selected separately, and breeding procedures involve mating males and females from various lines. Chicks are sexed at day-old age, and inaccuracies in sexing can lead to the mixing of cockerels with pullets. Birds identified with sexing errors should be removed before mating, and unwanted males must be culled early to reduce costs.

Selection of Individual Males and Females:

1. Assess the pedigree over several generations.
2. Evaluate the appearance and performance of the individual and its siblings.
3. Judge future potential based on the appearance and performance of progeny, typically sons and daughters.

Meat Type Lines:

- Since there is a high correlation between the weight of meat-type parents and their broiler offspring at eight weeks of age, selection should ideally occur at this age. Males require more selection pressure than females. Because the correlation between parent weight at sexual maturity and broiler offspring weight is low, selecting birds on a weight basis after eight weeks or at sexual maturity is challenging. Even if the same percentage of birds is culled from each sex, males face higher selection pressure because fewer males are needed than females.

Selection of Males:

- Initially, a higher percentage of males is retained to account for depletion due to mortality and culling. Typically, 60% of males are retained at day-old age, 40% at eight weeks, and 12-15% at sexual maturity. At eight weeks, at least 15% of birds should be weighed individually to represent the entire flock. Record weights from heaviest to lightest. Based on the required percentage, determine the number of males to retain, starting with the heaviest. The weight at which the retention percentage is reached becomes the minimum selection weight. Discard males below this weight. Weigh samples in every pen to account for variation.

Selection of Females:

- Females undergo less selection pressure than males, with about 80% retained at eight weeks. Body weight is less emphasized in selecting females. Selection may follow the same procedure as for males, but typically involves general appearance, body condition, conformation, and moulting pattern.

By integrating these methods, breeders can develop an effective and efficient breeding programme that enhances desired traits in their bird populations.

Conclusion

Poultry reproduction and breeding involve a complex interaction of anatomical, physiological, and genetic factors. The reproductive systems of chickens are specialized for efficient egg production and fertilization. Breeding systems aim to optimize genetic traits through methods like inbreeding, line breeding, outbreeding, and crossbreeding. Selection of breeding birds considers individual performance, family traits, and pedigree information, using techniques such as individual, family, and pedigree selection, as well as phenotypic value-based approaches like the tandem method, independent culling levels, and selection index. Mating methods, including individual mating, pen mating, flock mating, and artificial insemination, provide control and flexibility. Successful breeding programs combine genetic knowledge, selection techniques, and mating methods to improve production traits, maintain genetic diversity, and ensure poultry health, contributing to efficient and sustainable farming.

References

Bradley, OC (1960) The Structure of the Fowl, 4th Edition, Tom Grahame ed, Oliver and Boyd, London, UK. egyankosh,systems of poultry breeding.

Farhan sheikh,(2022) Breeding of poultry, the poultry Guide, Chicken.

North, Mack O and Bell, D (1990) Commercial Chicken Production Manual, 4th Edition, AVI Publishing Company, USA.

Parkhurst, CR and Mountney, GJ (1988) Poultry Meat and Egg Production, Von Nostrand Reinbold Company, Melbourne, Australia.

CHAPTER 7

BACTERIAL DISEASES IN POULTRY

[1*]Nayanmoni Konwar and [2]Sikhamoni Haloi

[1*] Assistant Professor, Department of Veterinary Microbiology, Institute of Veterinary Science & Animal Husbandry, SOA, DU, Bhubaneswar
[2] Assistant Professor, Department of Animal Nutrition, Institute of Veterinary Science & Animal Husbandry, SOA DU, Bhubaneswar, Odisha
*Corresponding author -nayanwork58@gmail.com

Abstract

Bacterial diseases causes great havoc in poultry farms. The diseases can affect both young chicks and adults and reduces their productivity. Bacterial diseases like Colibacillosis, Salmonellosis, Infectious coryza can cause high mortality in birds in short span of time. Antibiotic treatments are not always beneficial in case of these diseases. Proper diagnostic and preventive measures will help to curb the economic losses of the farmers.

Keywords: Poultry, Bacterial disease, Clinical symptoms, Diagnosis, Treatment

Introduction

Poultry industry has been playing a major role in developing India's economy and also contributes towards meeting the nutritional requirements of its citizens by egg and meat production. The Indian poultry market size has reached INR 1,905.3 billion in 2022 (Indian poultry market report, 2023-24). With the increase in intensive farming there is also high rise in several diseases of poultry. Bacterial diseases in poultry causes great harm to the farmers as there is huge mortality in birds. Some of the diseases are endemic in India and these disease causes great havoc in the farms. Several measures has been taken by the policy makers as well as the farmers to prevent the occurrence of the disease like vaccination of the birds, maintaining proper biosecurity in farms, following various treatment protocols. But inspite, of all these approaches bacterial disease are still causing mortality in birds. Through this chapter we are trying to provide information regarding the bacterial diseases which are very common. Information like the causative agents, clinical symptoms, prevention and control are provided in details which will be helpful for the students as well the farmers.

Diseases Caused by Salmonella in Birds

Salmonella is a gram-negative, non-spore-forming bacillus member of the Enterobacteriaceae family. Salmonellosis in poultry is mainly associated with non-typhoidal Salmonella (NTS) serotypes which includes serotypes *S. Typhimurium, S. Enteritidis, S. Heidelberg,* and *S. Newport*. The diseases caused by Salmonella in poultry are the following:

1. Pulloram disease

Pulloram disease mostly affects the newly hatched chicks and is responsible for considerable economic losses in the poultry industry worldwide. Historically the disease is also known as Bacillary White diarrhoea.

Etiology: The disease is caused by *Salmonella enterica enterica serovar Gallinarum biovar Pullorum* (*S Pullorum*). The bacteria is gram-negative non-sporogenic rods. They are non-motile under normal conditions and exhibit host specificity for poultry.

Host: The disease is mainly seen in newly hatched chicks. Poultry, turkeys, quails, ducks are susceptible to this disease. The young chicks die between 2-7days old age and the mortality may go up to 100%.

Transmission

The disease is transmitted vertically (transovarian) through infected eggs and also via direct and indirect contact with infected birds or contaminated feed, water or litter. The adult birds acts as asymptomatic carrier.

Clinical symptoms

The disease is mostly seen in newly hatched chicks and very rarely in adults and growers.In the hatchery there will be dead in shell in more numbers. The birds will huddled near a heat source, there will be loss of appetite, depression and whitish diarrhoea. The survived birds will be smaller in size and will act as subclinical carriers.

Pathological lesions

In the newly hatched chicks there is no marked gross lesions. Slight haemorrhagic streaks on liver can be observed. In grown chicks there is necrotic spots in the liver, raised lungs and whitish spots on heart and spleen. There will be presence of unabsorbed egg -yolk in young chicks and there is presence of semi-solid, cheesy material in the caecum. In adult birds there will be presence of cystic lesions in the ova. Pericarditis may be present.

Diagnosis

- Mortality in hatchery or brooder must be suspected for salmonellosis
- Isolation of the organism in suitable media
- Whole blood agglutination test or tube agglutination test can be done
- Molecular detection

Prevention and control

- Chicks should be brought from disease free flocks
- The hatcheries should be properly disinfected before putting a new batch
- The carrier birds should be detected and removed from the flock
- Proper biosecurity measures should be followed

Treatment

- Treatment doesnot prevent the affectected birds from becoming carriers
- Antibiotics like Trimethoprim, ampicillin, sulphadiazine are beneficial

2. Fowl typhoid

Fowl typhoid is a septicaemic disease which is mostly present in some regions of India like Assam, Bengal, Uttar Pradesh and Bihar. The diseases causes much economic losses to poultry farmers and is one of significant cause of mortality in birds.

Etiology: The disease is caused by *Salmonella gallinarum* which is a gram negative, non -motile, host adapted bacteria and it is antigenically similar to *S. pullorum.*

Hosts: The disease may occur in birds of all age group but it is more common in birds of 3-6 month age.

Transmission: The organism is transmitted through eggs and also by horizontal transmission with contaminated feed and water. The disease has high tendency to spread among matured birds and mortality in young birds is similar to that in pullorum disease.

Clinical symptoms: Affected bird shows dullness, loss of appetite, fever and yellowish diarrhoea. The wattles and comb are cyanotic. Death occurs from 2nd day of hatch and gradually declines by 8th day post hatch. Mortality varies from 10-50% and may last up to 5 months of age.

Pathological lesion: In case of acute disease there is enlargement of liver and spleen and characteristic copper colour liver is observed. Discoloured and deformed ova is seen in affected birds. There is presence of small, grey nodules in heart and greyish consolidation in lungs.

Diagnosis

- Diagnosis is similar to pullorum disease but differential diagnosis should be done.

Prevention and control

- Identification and elimination of infected breeding flocks
- Strict biosecurity measures should be taken
- Vaccination is not full proof in case of this disease

Treatment

- Antibiotic treatment can be provided to the affected flock.

3. Paratyphoid

Fowl paratyphoid is an acute or chronic disease infecting domestic fowl and many other avian or mammalian species. It is caused by non host adapted Salmonella serotypes which are motile. The highest morbidity and death rates are usually observed during the first 2 weeks after hatching. The chickens are drowsy, with eyes closed, ruffled feathers and grouped near the sources of heat.

Paratyphoid infections are of public health importance as there may be contamination due to mishandling of poultry products. Paratyphoids are usually present in the intestinal tract of poultry without causing any disease; however, a few serotypes have the ability to cause disease in poultry, particularly when they are vertically transmitted.

Etiology: Paratyphoid infections is cause by any one of the many non–host-adapted salmonellae. *Salmonella enterica serotype Typhimurium, S enterica serotype Enteritidis, S enterica serotype Kentucky*, and *S enterica serotype Heidelberg* are among the most common Salmonella infections in poultry.

Transmission

Transmission of the disease usually occurs horizontally from infected birds to uninfected ones, from contaminated environments, or infected rodents. Transovarian transmission is observed in case of *S Enteritidis* and *S enterica serotype Arizonae* but in case of most serotypes infection to progeny from infected breeders is mainly through fecal contamination of the eggshell. Infected birds remain carriers.

Clinical symptoms

Clinical signs are mostly observed in young birds and death in young birds is most often limited to the first few weeks after hatch. There is no any distinctive clinical sign. In the affected birds there will be depression, poor growth, weakness,diarrhoea, dehydration, decreased feed consumption and decrease in egg production in case of layer birds.

Pathological lesions

In case of acute infection in young birds post mortem findings will reveal an enlarged liver with focal necrosis, unabsorbed yolk sac will be present, presence of enteritis with necrotic lesions in the mucosa and cecal cores. Occasionally there will be localized infection in the eye or synovial tissues. In acute cases the birds may die without any lesions but mainly due to septicemia.

Diagnosis

- Confirmation is based on isolation and identification of Salmonella spp from liver, spleen, intestine, or heart blood.
- Routine microbiological screening of liver/spleen/intestinal pools from post mortem submissions to laboratories is strongly recommended.
- Specific ELISA-based test kits are available for assaying for *S. enteritidis* antibody.

Treatment

Treatment with antimicrobials is not recommended as antimicrobial agents will help in reducing mortality rate but it cannot eliminate flock infection. Excessive use of antibiotics may contribute to emergence of antimicrobial resistance.

Prevention

- Screening of the birds should be done with appropriate biosecurity procedures.
- Proper sanitation should be maintained in the hatchery and poultry houses
- Vectors such as wild birds, rodents, pets and flies should be eliminated
- Feed management should be done
- Vaccination should be done with live modified *S. typhimurium* vaccines alone or in combination with inactivated *S. enteritidis* emulsion boosters at point of lay.

Diseases due to Escherichia Coli

Coliform infections are important disease of poultry. Numerous conditions can be caused by colibacillosis such as cellulitis, airsacculitis, salpingitis, peritonitis, acute lethal septicemia and subacute pericarditis. While swollen head syndrome and omphalitis (infection of the yolk sac) are common diseases in newborn chicks, peritonitis and salpingitis are common in laying hens.

1. Colisepticaemia and airsacculitis

Etiology: The disease condition is caused by *Escherichia coli b*elonging to the serogroup O1, O2, O35 and O78 strains. The bacteria is gram negative rods, motile and can grow under both aerobic or anaerobic conditions.

Transmission

The bacteria enter into the birds through various manner but of the main route of infection is considered to be the respiratory tract, abrasions in skin, cloaca, damaged intestinal mucosa and navel. Through these portals of entries, *E coli* can replicate locally and then goes into the bloodstream to cause colisepticemia, which may progress from acute septicemia to death.

Clinical symptoms

In young chicks there is diarrhoea and pasty vent. The affected grower birds shows symptoms like depression, loss of appetite, dyspnoea and sneezing.

Pathological Lesions

The most important characteristic lesion is the presence of milky fluid in the pericardium. The membranes of air sac is thickened and cloudy in appearance. The liver is enlarged and a thin fibrinous exudate is present over the liver.

Diagnosis

- Clinical symptoms like death in young birds with pasty vent
- Isolation of the organism in media specific for *E.coli*
- Biochemical test
- ELISA

2. Coligranuloma (Hjarre's disease)

This disease is caused by *E.coli* belonging to the serogroup O4,O8 and O16.It is a disease of adult chicken and turkey and characterized by presence of nodular granuloma in the liver, mesentary and intestine. The nodules are size of millet seeds or sometimes larger and they are also present in caecum, kidney, lungs and gizzard. Differential diagnosis should be made from avian tuberculosis.

3. Omphalitis

This is a very important disease in newly hatched chicks. Young chicks up to 1 or 2 weeks of age are affected. The disease mainly occur due unhygienic condition of the hatchery, temperature fluctuation in the hatchery, production of weak chicks which are susceptible to infection. The affected chicks are mostly huddled together and there is depression similar to paratyphoid. In the affected birds the abdomen is distended as the yolk sac is unabsorbed and the contents are cheesy and foul smelling. The umbilicus of the infected chick is inflammed and red and there may be presence of pericarditis in chicks more than 4 days old.

Diagnosis of the disease can be done with clinical symptoms, post mortem lesions, isolation of the organism, various biochemical test and PCR.

Diseases Caused by Mycoplasma in Birds

1. Chronic Respiratory Disease in poultry

Etiology: *Mycoplasma gallisepticum* causes chronic respiratory disease in chickens and infectious sinusitis in turkeys.

Host: The disease mostly affects 4-10 weeks old birds especially broilers and Turkeys. Very occasionally it may infect older birds. Other avian species like ducks, geese, guinea fowl, quails and pigeon are also infected.

Transmission

The transmission of mycoplasmosis is both by vertical and horizontal routes. The infected parents can pass the disease to their progeny by the vertical route from infected parent flocks to progeny. Horizontal transmission mostly occurs by direct contact between clinically affected or recovered carriers and susceptible flocks. Indirect infection occurs through contact with contaminated equipment, feed bags, and personnel. Susceptible flocks may contract the disease via wild birds and rats. Mycoplasmas cannot survive outside the host for longer than 24 hours.

Clinical signs

The main clinical sign is the involvement upper respiratory tract in chickens. The starting signs of the disease are sneezing, coughing, respiratory distress and gargling sounds during respiration. There is presence of nasal discharge and conjunctivitis with frothiness in the eyes. Turkeys are highly susceptible to this disease and infection results in swelling of the infraorbital sinuses and is thus it is called "Infectious sinusitis". Feed efficiency and weight gains are reduced. Due to the presence of airsacculitis the commercial broiler markets suffer high loss due to condemnation during processing. In laying flocks, there may be a chronic increase in mortality and a decrease in the overall production rate. Uncomplicated *M gallisepticum* infections in chickens result in relatively mild catarrhal sinusitis, tracheitis, and airsacculitis. Turkeys develop severe mucopurulent sinusitis and varying degrees of tracheitis and airsacculitis. Morbidity is high and mortality is low in uncomplicated cases.

Diagnosis

- The lesions of air sac at postmortem examination are very important.
- Diagnosis is based on isolation and identification of the pathogen and on flock testing using the serum plate agglutination test.
- Several PCR-based methods for the detection of the organism in clinical samples and for identification of isolates have been published (Anon., 2008c).
- Haemagglutination inhibition test
- ELISA

Prevention

- Eggs used for hatching should be dipped in a tylosin solution to eliminate the pathogen.The eggs are warmed at 37.8°C for 2-3 hours and then dipped in cold watery solution of 0.04%to 0.1% tylosin Tartarate or erythromycin. The eggs are dipped in the solution for 10-30 min. This method helps in entry of antibiotics into the eggs which enables to kill the organisms if present.
- Modified live vaccines and bacterins are available.

2. Infectious synovitis

Etiology: *Mycoplasma synoviae is* the cause of infectious synovitis in chickens and turkeys, is transmitted mainly by aerosols.

Host: The disease is commonly seen in growers of 4-12 weeks age in chickens and in turkeys the disease mostly occurs in 10-20 weeks of age. The organism is transmitted by egg and horizontal transmission is also observed.

Clinical symptoms

The birds which are under stress are more likely to be clinically affected. The first clinical sign observed is pale bluish head and presence of lameness. In some birds slight raling sound might be present due to respiratory infection. The severely affected birds are depressed with swollen hocks and footpads. Sternal bursitis may also be observed. Synovitis, arthritis and respiratory signs are the main clinical features.

Diagnosis

- Diagnosis of the disease can be done by isolation and identification of the pathogen or positive serological tests.
- Serum aggluniation test, ELISA can be done
- PCR-based detection methods are also available.

Tetracycline medication of the feed is used for treatment and control. Eradication is possible through the development of specific-pathogen-free flocks

3. Mycoplasma meleagridis

Mycoplasma meleagridis causes diseases in breeder turkeys and the organism is transmitted through the eggs (transovarian). The disease causes airsacculitis in the progeny as well as decreased hatchability with poor growth and skeletal abnormalities in progeny. *M meleagridis* is strongly host-specific for turkeys and the infection primarily occurs through vertical (egg) transmission with rates that vary over the laying season.

Prevention

- It is essential to purchase parent and commercial stock from known mycoplasma-free breeder flocks.
- Strict biosecurity will prevent lateral introduction of infection.

Diseases Caused by Clostridium Organism

1. Necrotic enteritis in poultry

Necrotic enteritis is a disease which is characterized by acute enterotoxaemia. The disease mainly effects broilers and turkeys and is characterized by sudden increased in flock mortality. It is an acute enterotoxaemia characterized by sudden onset and high mortality. Confluent necrotic areas in the mucosa of the small intestine are found post mortem. Dietary changes, intestinal hypomotility and mucosal damage caused by coccidia and other enteric pathogens predispose to the development of disease.

Etiology: The disease is caused by *Clostridum perfringens type A* which is a gram-positive, obligate, anaerobic bacterium. The bacteria can be usually isolated on blood agar, incubated anaerobically at 37°C (98.6°F), on which it produces a double zone of hemolysis. Toxins produced by the bacteria causes extensive damages to the small intestine and liver.

Host: Broilers, breeders, turkeys, pullets and ducks are highly susceptible to this disease. The birds between 2-5 weeks age are most commonly affected.

Clinical symptoms

The main clinical sign of this disease is sudden increase in mortality of birds. The affected birds will be dull, ruffled feathered and there will be diarrhoea. There is drooling of saliva in some cases. In acute cases the birds may die in 1-3 days post infection.

Pathological lesion

The small intestine is mostly infected and there is necrosis of mucosal layer which turns yellowish colour and dry in appearance. The liver becomes small and mottled and gall bladder is distended with bile. There is atrophy of the breast muscle, spleen and testes. Cholangiohepatitis may also be an associated disease with necrotic enteritis.

Diagnosis

- Isolation can be done in Blood Agar (round ,smooth ,convex , 2-3mm in diameter) at 37 C , in presence of 2-10% CO
- Nagler's reaction- *C. perfringens* produces an opalescence surrounding the streak line on egg yolk agar (EYA) or media containing 20% human serum due to **lecithinase activity of alpha toxin**.

- Robert'son cooked meat broth (RCM) : meat particle turns into pink

Prevention and control

- Care should be taken to prevent the occurrence of coccidiosis
- Excessive feeding of animal by-products, rye, fish meal, wheat, barley in the diet should be avoided
- Administration of probiotics or competitive cultures should be done
- Treatment with antimicrobial-medicated drinking water
- Due to the ubiquitous nature of *C perfringens,* necrotic enteritis can be controlled by preventing coccidiosis, which act as the main predisposing factor for *C perfringens* overgrowth and infection.

Treatment

- Antibiotics should be administered in drinking water to treat the infected birds. Antimicrobials like bacitracin (52.8 mg/L for 5–7 days), penicillin (400,000 U/L for 5 days), and lincomycin (17 mg/L for 7 days) are most commonly used
- The weak birds should be removed promptly because they can be a source of toxicosis or infection as a result of cannibalism.

2. Botulism (Limberneck)

Botulism occurs due to ingestion of the exotoxin generated by *Clostridium botulinum,* neurotoxin-producing bacteria. It is found all across the world and affects a large range of birds and mammals. Leg weakness and paresis are the first clinical symptoms, which proceed to a flaccid paralysis that affects the nictitating membranes after the legs.

Etiology

Clostridium botulinum is a gram positive,spore forming anaerobic organism. The botulinum toxins are neurotoxins in nature and have affinity for motor neurons. The toxin attaches itself irreversibly to the presynaptic membrane of cholinergic nerve terminals after being absorbed. It penetrates the cell and causes flaccid paralysis by preventing the release of acetylcholine across neuromuscular junctions. Respiratory and cardiac failure leads to death.

Host: The free ranging poultry are usually affected by this disease as in free range they sometime eat putrefied feed when the bacteria is producing toxin.

Clinical symptoms

The incubation period of the disease is from 2 hours to 2 weeks. The appearance of clinical symptom depends on the amount of neurotoxin toxin consumed by the bird.

There is development of leg weakness and paresis that progress to flaccid paralysis of the legs, wings, neck and nictitating membranes. In birds the condition is called Limber neck due to the neck paralysis typically observed in affected birds. In broiler chickens there is a ruffled or quivering feathers, feathers that are easily pulled out and labored breathing.

Birds that are severely afflicted lie on the ground in ventral recumbency, with their necks outstretched and their eyes closed or partially closed. The flaccid paralysis that develops prevents them from raising their eyes and from being able to lift or support their neck.

Lesions

(Botulinum neurotoxin) BoNTs do not cause lesions in affected birds, but maggots or other invertebrates may be found in the crop. Carcasses of dead birds are usually in good condition and do not exhibit sign of a chronic or debilitating disease.

Diagnosis of Botulism in Poultry

- Detection of toxin
- Identification of toxin-producing clostridia
- Clinical symptoms in the affeted birds
- Mouse lethality assay.
- ELISA and mass spectrometry
- Molecular diagnosis

Control and Prevention of Botulism in Poultry

- Proper disposal of dead birds should be done
- Cleaning and disinfection of the farm should be done in regular basis
- Biosecurity measures should be very efficient
- Vaccination can be done with a toxoid or antitoxin

Treatment

- Supportive care should be given to the affected birds
- Beta-lactam antimicrobials have been proved beneficial for treting the disease during outbreaks.
- Botulism antitoxins specific to the toxin type can help in neutralizing the circulating botulinum toxins in birds.

3. Ulcerative enteritis (Quail disease)

The disease was first identified in bobwhite quail (*Colinus virginianus*). Ulcerative enteritis also affects chickens, turkeys, pheasants, grouse, and other gallinaceous birds. There have also been reports of the disease in pigeons, psittacine birds and Japanese quail (*Coturnix coturnix japonica*).

Etiology: The disease is caused by *Clostridium colinum.* It is an anaerobic, fastidious gram-positive, spore-forming, slightly curved rod, ~1 × 3–4 mcm wide, with subterminal, oval spores.

Transmission: Flies that eat contaminated excrement or recovered carrier birds can spread the infection. The bacteria are excreted by infected birds in their droppings.

Clinical symptoms

In affected bobwhite quail, ulcerative enteritis can cause rapid death in as little as two to three days, with no symptoms or weight reduction. Hemorrhagic enteritis of the duodenum is an example of an acute lesion. Anorexia coexists with a less severe course of the disease in hens and other game animals. Similar symptoms to those of coccidiosis include diarrhea, ruffled feathers, humped backs, depression and occasionally bloody or watery white droppings, especially in quail throughout the prolonged course of the disease. In 2–3 weeks, chickens heal, and mortality rarely rises above 10%.

Lesions

In early stage of the disease there is presence of small, round ulcers surrounded by hemorrhages in the small intestine, caeca, and upper large intestine. The small ulcers later coalesce to form larger ulcers, sometimes the ulcers perforates the intestinal layer producing local or diffuse peritonitis.

Diagnosis

- Postmortem lesions should be examined
- Demonstration of bacilli from the necrotic tissues
- Culture of the organism
- Fluorescent antibody test
- Molecular diagnosis by PCR

Prevention and control

- Bacitracin in feed is used as a preventive against the disease in quail.
- Streptomycin and furazolidone are effective treatments.
- Streptomycin (0.006%) and furazolidone (0.02%) are added to feed effective to treat the disease.
- The introduction of new birds into existing flocks should be avoided.
- High population density is a predisposing factor. The use of cages is recommended in quail breeding.
- Sick and dead birds should be removed immediately.
- The farm should be totally cleaned before introducing new batch of birds

Others

1. Fowl Cholera or Pasteurellosis

Fowl cholera is a septicemic disease that affects both farmed and wild birds. This is a extremely contagious disease, affects especially in large industrial-type chicken complexes, results in substantial morbidity and death as well as significant economic losses. Although it usually manifests as an acute illness, certain outbreaks can potentially result in chronic infections.

Etiology: *Pasteurella multocida,* is a gram-negative, non-motile, non-spore-forming bacillus. In freshly isolated cultures the bacteria has a bipolar appearance when stained with Wright stain.

Host: Adult birds and old chickens are more susceptible to this disease. In parental flocks, the males are far more susceptible than hens. Besides chickens, the disease also infects turkeys, ducks, geese, raptors and canaries, particularly Turkeys as they are highly sensitive with mortality ranging to 65%.

Transmission

Infection mainly occurs following direct contact between susceptible birds and clinically affected or recovered carriers. Environmental contamination, rodents and wild birds may act as sources of indirect infection. The disease can also be introduced into the farm from contaminated feed bags, equipment and the clothing of personnel.

Clinical symptoms

In a flock the morbidity and mortality rates depends on the pathogenicity of the strain and the susceptibility of the flock. In acute cases there will large number of mortality in the flocks without much clinical symptoms. Newly introduced infections may result in up to 10% mortality. In more protracted cases there is presence of listlessness, anorexia, mucoid discharge from the mouth, ruffled feathers, diarrhea, and increased respiratory rate. Pneumonia is particularly common in turkeys.

In chronic case of fowl cholera there is localized infections of the sternal bursae, wattles, joints, tendon sheaths and footpads, which often are swollen because of accumulated caseous or fibrinous exudate. Lameness, as well as exudative conjunctivitis and pharyngitis may occur in some cases. Torticollis may also occur when the meninges, middle ear, or cranial bones are infected.

Pathological lesions

In case of acute infection there is enlargement of spleen and liver with punctate hemorrhages of the viscera including the heart. In subacute cases there is presence of gray granulomatous foci in the liver. In chronic case there is presence of caseous cellulitis of the wattles and seropurulent arthritis.

Diagnosis

- Isolate and identification of *P. multocida* from specimens of heart blood, liver and spleen.
- In acute cases bipolar organisms can be demonstrated in direct smears prepared from heart blood
- PCR assay
- Serological testing - rapid whole blood agglutination
- Serum plate agglutination
- Agar diffusion tests and ELISA.

Treatment

Tetracycline incorporated into feed at a level of 200 - 400 g/ton or in water at 250 - 500 mg/l will suppress clinical signs and reduce mortality.

Prevention

- Strict biosecurity protocols are required to stop the spread of infection.
- Elimination of rodents is essential for decreasing flocks' exposure to *P. multocida*.
- In endemic areas the flocks should be immunized.

Vaccination should be done at 10 and 14 weeks during the growth period with live-attenuated *P. multocida* vaccinations (CU; PM-1; PM-9 strains)

2. Avian Spirochaetosis

Avian spirochetosis is an acute bacterial infection which is transmitted by tick to a wide range of birds.

Etiology: The disease is caused by *Borrelia anserina,* an actively motile spirochete, ~0.2–0.3 mcm × 8–20 mcm, and consists of 5–8 loosely arranged coils. The disease is mainly prevalent in tropical and subtropical regions where the disease is endemic-Pakistan, India, Iran, Brazil, the US, Algeria, Sudan, Ethiopia, Nigeria and Tanzania.

The vector of avian spirochaetosis is Argas persicus, the fowl tick; however, other Argas spp, eg, *A sanchezi*, transmit the bacteria as well.

Clinical symptoms

In some birds the infection may remain subclinical and incase of infected birds the clinical signs include listlessness, somnolence, moderate to marked shivering and increased thirst. Ruffled feathers, anemia, and pale combs are observed along with in appetence, which leads to drastic weight loss.

During the early stages of the disease, the affected birds usually develop a green or yellow diarrhea with increased urates. The course of the disease is 1–2 weeks. In case of layers the egg production may be decreased by 5%–10%, with a higher number of small eggs.

Diagnosis

- Diagnosis can be confirmed by demonstration of the spirochaetes in buffy coat smears using dark-field microscopy.
- Blood or tissue smears can also be examined using immunofluorescence.
- Giemsa - stained smears or silver impregnation techniques are used to demonstrate the borreliae in tissues.
- Borrelia will grow on Barbour-Stoenner-Kelly medium but loses virulence after 12 passages.
- The organisms are usually isolated by inoculating embryonated eggs or young chicks with infected blood or homogenized tissues.
- PCR techniques have been applied for the characterization of *B. anserina* strains .

Treatment

- Treatment with antibiotics is effective. Inactivated vaccines and tick eradication are the main control measures.
- Penicillin derivatives, aminoglycosides, tetracyclines and tylosin are also effective. Tiamulin is labeled for treatment of spirochetosis in poultry outside of the US

Control

- Controlling the tick population is the main effective measure

3. Infectious Coryza

Infectious coryza is an acute respiratory disease of chickens and it has been recognized since the 1930s. The disease is worldwide in occurence and causes economic losses due to an increased number of culls and a marked drop in egg production, particularly on multi-aged farms.

Etiology: The causative agent of infectious coryza is *Avibacterium paragallinarum*,also known as H*aemophilus paragallinarum.* It a gram-negative, pleomorphic, nonmotile, catalase-negative, microaerophilic rod that requires nicotinamide adenine dinucleotide (V-factor) for culture.

Transmission

Birds with chronic illness or those that are healthy carriers may serve as a source of infection for *A paragallinarum*. Chickens of all ages can become infected, but the likelihood of being infected increases as they get older. The time from exposure to onset of symptoms is 1-3 days, and the usual length of illness is 2-3 weeks. Transmission occurs through direct contact, airborne droplets and contamination of drinking water. Transmission does not happen through eggs. "All-in/all-out" management has effectively eliminated infectious coryza from numerous commercial poultry operations in the US.

Clinical symptoms

Infectious coryza is a disease that affects both grower and layers. The most common clinical signs are serous nasal discharge, facial swelling, lacrimation, anorexia and diarrhea. In severe cases there is extreme swelling of one or both infraorbital sinuses and edema of the surrounding tissues that may prevent the eyes from fully opening. Due to decreased feed and water consumption there is less growth in young stock and reduced egg production in laying flocks. In adult birds, especially males, the edema may extend to the intermandibular space and wattles. The swelling usually abates in 10–14 days; however, if secondary infection occurs it can persist for months. There may be varying degrees of rales depending on the extent of infection.

Diagnosis

- Isolation of organism from infected birds
- Clinical symptoms
- Polymerase chain reaction is a sensitive method for bacterial identification.
- Animal inoculation techniques

Control and Treatment of Infectious Coryza in Chickens

- Sound management practices and vaccination can help prevent infection in birds
- Antibiotic treatment along with supportive care of infected birds helps in recovery
- Prevention of the disease is important for control of infectious coryza
- All in all out concept should be followed in the farm
- Vaccination should be done before 4 weeks of age

Treatment

- Immediate antibiotic treatment should be made available if the disease is spreading.
- Erythromycin and oxytetracycline are usually effective.
- Additionally, several newer-generation antimicrobials (eg: fluoroquinolones, macrolides) are effective against infectious coryza.
- Various sulfonamide drug including trimethoprim-sulfamethoxazole, and other drug combinations have been successful for treatment.

4. Avian Tuberculosis

Avian tuberculosis is one of the most important diseases that affect both domestic and pet birds. The disease is most often caused by *Mycobacterium avium* belonging to serotypes 1, 2, 3, and 6.

Transmission

The mycobacteria can survive in the environment for several months and can infect the birds through contaminated water and soil. Places with high population density and poor sanitation and hygiene can act as a source of infection. The free roaming birds are highly susceptible to the diseases as well in spreading the disease. *M avium* is ingested by the birds, although infection may be transmitted via aerosol.

Clinical symptom

The incubation period of the disease is very long and can prolong for weeks to months. The disease is more prevalent in adult birds and the losses are experienced more in older stocks of age group 18–20 months. The disease process can be divided into three phases: Latency, lesion formation and cachexia period.

The growth of enormous tubercles with a high number of bacilli occurs during cachexia. A third type of infection is frequently observed as a nontuberculous one, primarily seen in finches, canaries and psittacines. In the traditional form of infection, tubercles or granulomas grow in many organs; in the second form, lesions in the intestinal system are the manifestation.

Lesions

The principal lesions of tuberculosis are the presence of greyish-white to greyish-yellow nodules in intestine of birds.The spleen is knobbly in appearance. There is presence of caseous lesions in the liver and spleen and enlargement of the organs. Nodules are firm but can be incised easily since mineralization is rare in avian TB. The bone marrow of the long bones frequently contains tubercular nodules. Some exotic bird species may have lesions in the liver and spleen without intestinal involvement.

Diagnosis

- The diagnosis of disease is based on clinical signs, postmortem gross lesions, and by demonstrating the acid-fast bacilli in crushed lesions using microscopy, which is sufficient for a positive diagnosis
- Culturing the organism in suitable media, namely, Dorset's or Herrold's egg yolk medium, Lowenstein-Jensen mediume
- For growth of *M. avium*, media containing whole egg or egg yolk is desirable and the incubation temperature should be 37–40 C.
- ELISA

- Agglutination test
- Molecular assay

5. Avian Chlamydiosis

Avian chlamydiosis is a disease that may cause subclinical infection or as an acute, subacute, or chronic illness in both wild and domestic birds, presenting with respiratory, digestive, or systemic symptoms.

Etiology: Avian chlamydiosis is caused by *Chlamydia psittaci*, an obligate intracellular bacterium. Currently 11 species are recognised under the genus Chlamydia and among them *C. psittaci, C. avium* and *C. gallinacea* have been isolated from birds.

Transmission

The primary mode of spreading is through either the fecal-oral route or through inhalation. Respiratory discharge or feces from infected birds carry elementary bodies that can survive drying and remain contagious for months if shielded by organic material like litter and feces. Airborne particles and dust also can carry the organism.

Clinical symptoms

The respiratory system is mostly infected in turkeys and chickens. There is nasal and ocular discharge, conjunctivitis, sinusitis, green to yellow-green droppings, fever, inactivity, ruffled feathers, weakness, inappetence, weight loss and drop in egg production. The disease is more severe in young birds. There is anemia and leukocytosis with heterophilia and monocytosis. Plasma bile acids, AST, LDH, and uric acid may be increased. Enlarged liver and spleen may be observed on radiography.

Lesions

Birds infected with *C. psittaci* show a range of lesions, which can encompass nasal inflammation, lung congestion, fibrinous pneumonia, inflammation of the air sacs, an enlarged spleen (which may appear mottled or discolored), and an enlarged liver with multiple areas of liver damage. Additionally, these birds may experience fibrinous inflammation around the liver, inflammation of the heart lining, inflammation of the abdominal lining and swelling of blood vessels. They might also present with inflammation of the intestine and the lining of the eye. Birds that are not showing symptoms often do not have any visible signs of infection.

Diagnosis

- Isolation of *C psittaci* in chick embryo or cell cultures
- PCR is a highly sensitive toolfor diagnosis of the disease
- Immunohistochemistry techniques are used for detection of the organism

- Hematological examination
- Various serological test

Prevention and control

- Proper biosecurity practices are to be undertaken to control the introduction and spread of chlamydiae
- All the new birds should be kept in quarantine
- Exposure to wild birds should be prevented
- Cross-contamination should be minimized by decreasing people entering the farm
- The affected birds should be isolated and treatment should be provided
- Disinfection of the entire farm and the premises
- The feeds should be uncontaminated

Treatment

- Tetracycline is the choice of drug for the treatment of this disease

Zoonotic importance of Avian Chlamydiosis

Avian Chlamydiosis is a zoonotic disease that can impact individuals following contact with airborne particles from the digestive or respiratory systems of infected birds, whether they are alive or deceased, or through the handling of bird tissues (such as those from a slaughterhouse), their bedding, or other materials. The illness in humans typically arises from contact with pet parrots and can happen even with just a short interaction with an infected bird. Other individuals at risk include those who work closely with birds, such as pigeon enthusiasts, veterinarians, farmers, those who care for injured wildlife, zookeepers, and staff in slaughtering and processing facilities or hatcheries. The spread of Chlamydiosis from birds to workers in the poultry industry is likely underreported. When dealing with live or deceased birds that are infected, it's important to take precautions to avoid exposure. This includes wearing protective gear such as a dust mask and a plastic face shield or goggles, gloves, using a detergent disinfectant to clean feathers and wearing a hood that has been exposed to air to prevent dust from entering the respiratory system. The illness in humans can range from showing no symptoms at all to experiencing symptoms similar to the flu, including respiratory issues like pneumonia. In some cases, it can lead to more serious conditions such as endocarditis, myocarditis, hepatitis, and encephalitis. Individuals with weakened immune systems are more likely to develop symptoms of the disease.

Conclusion

In this chapter information regarding various bacterial diseases of poultry has been well described. The etiology of the diseases, their clinical symptoms, diagnosis, treatment, prevention and control has been provided. When there is disease outbreak the live birds which are showing symptoms should be separated and samples should be sent to poultry diagnostic laboratory to identify the disease and proper control measures should be taken. These will help in control the spread of the disease in the farms and there will be less mortality and losses.

References

Berry, J. G. (2004). Bacterial Diseases of Poultry. Division of Agricultural Sciences and Natural Resources. Oklahoma State University

Blackall P. J. (1999). Infectious coryza: overview of the disease and new diagnostic options. Clinical microbiology reviews, 12(4), 627–632.

Dhama, K., Mahendran, M., Tiwari, R., Dayal Singh, S., Kumar, D., Singh, S., & Sawant, P. M. (2011). Tuberculosis in Birds: Insights into the Mycobacterium avium Infections. Veterinary medicine international, 2011, 712369.

MSD veterinary manual

N. Kuria, J. (2024). Salmonellosis in Food and Companion Animals and Its Public Health Importance. IntechOpen. doi: 10.5772/intechopen.109324

OIE (2024). Manual of Diagnostic Tests and Vaccines for Terrestrial Animals,(13th ed.).

Shaji, S., Selvaraj, R. K., & Shanmugasundaram, R. (2023). Salmonella Infection in Poultry: A Review on the Pathogen and Control Strategies. Microorganisms, 11(11), 2814.

Shane, S. M., & Association, A. S. (2005). *Handbook on Poultry Diseases*.

CHAPTER 8

POULTRY VIRAL DISEASES

*[1]M. Bindu Pushpa, [2]Salur Srihitha, and [3]Uppari Anusha Rao

[1] Assistant Professor, Department of Veterinary Public Health, Institute of Veterinary Science & Animal Husbandry, SOA DU, Bhubaneswar, Odisha
[2] Assistant Professor, Department of Animal Nutrition, Institute of Veterinary Science & Animal Husbandry, SOA DU, Bhubaneswar, Odisha
[3] Assistant Professor, Department of Veterinary Parasitology, Institute of Veterinary Science & Animal Husbandry, SOA DU, Bhubaneswar, Odisha
*Corresponding author- bindupushpa1221@gmail.com

Abstract

Poultry viral diseases present substantial risks to the global poultry industry, endangering both economic stability and food security. These diseases, which affect a variety of poultry species, include Avian Influenza (AI), Infectious Bronchitis (IB), and Marek's Disease (MD) Newcastle Disease (ND). AI is highly contagious with severe respiratory symptoms, ND affects multiple systems leading to substantial losses, IB impacts the respiratory system and egg production, while MD causes tumors and immunosuppression in young chickens. Control relies on biosecurity measures, vaccination and surveillance. Vaccines for ND, IB, and MD are key, though challenges like efficacy and virus mutation persist. Essential biosecurity practices include farm hygiene, controlled bird movement, and proper disposal of dead birds. Early detection and surveillance are crucial for outbreak management and infection control.Advancements in diagnostics, vaccines and understanding viral pathogenesis are vital. Ongoing research and international collaboration are necessary to develop effective strategies to mitigate these diseases impacts on poultry health and the industry.

Keywords: Diagnostic techniques; Poultry viral diseases; Surveillance; Viral pathogenesis

Introduction

The Indian poultry industry, growing at 8.1% annually, faces significant challenges from viral diseases like Newcastle disease, Avian Influenza, Fowl Pox, and others, which impact economic progress. Epidemiological research, including genotyping and molecular characterization, is key to controlling these diseases. Strengthening biosecurity, hygiene, and vaccination efforts is essential to mitigating outbreaks, as they can severely affect poultry production and the broader economy.

NDV (Newcastle disease virus) or Avian pneumo-encephalitis or Ranikhet disease or Exotic newcastle disease (END)

Newcastle Disease (ND), also known as Ranikhet Disease, is a highly contagious and often fatal poultry disease, leading to high morbidity, mortality and reduced egg production. Chickens, especially young ones, are highly susceptible, while ducks may carry the virus with minimal or no symptoms.

Aetiology:

ND is caused by Avian Orthoavulavirus 1 (AOAV-1), also known as Newcastle Disease Virus (NDV), part of the Avulavirinae subfamily in the Paramyxoviridae family.

Pathotypes:

NDV strains are categorized by virulence and affected tissues:

1. Velogenic Strains: Highly fatal, causing severe symptoms (lethargy, inappetence, respiratory issues, diarrhea, neurological signs).
   - Viscerotropic Velogenic: Causes hemorrhagic intestinal lesions.
   - Neurotropic Velogenic: High mortality with respiratory and nervous symptoms.
2. Mesogenic Strains: Moderate virulence, causing respiratory and neurological signs; low mortality (<10%).
3. Lentogenic Strains: Mild, mainly respiratory symptoms; typically subclinical.
4. Asymptomatic Enteric Isolates: Subclinical intestinal infections.

Clinical Signs:

Birds may show depression, ruffled feathers, greenish-white diarrhea, neurological symptoms (torticollis, paralysis) and high mortality within 2-3 days. The incubation period is typically 3-6 days.

Necropsy Findings:

Findings include mucus in the trachea and hemorrhaging in the intestines, especially in the proventriculus

Diagnosis:

- Definitive: Virus isolation and laboratory characterization.
- Presumptive: Clinical signs and lesions.
- Laboratory Confirmation: Virus isolation in embryonated eggs.
- Serological Tests: Haemagglutination inhibition, ELISA, viral neutralization.
- Molecular Methods: RT-PCR and qPCR for higher sensitivity.

Avian influenza (AI)

Avian Influenza (AI) is a highly contagious viral disease in poultry, characterized by high morbidity and mortality, respiratory signs, depression, and reduced feed and water intake. It also leads to a decline in egg production in laying birds.

Aetiology:

- AI is caused by type A influenza viruses from the Orthomyxoviridae family, which have a segmented, single-stranded, negative-sense RNA genome within an envelope.
- AI viruses are classified into Low Pathogenic (LPAI) and Highly Pathogenic (HPAI) strains. HPAIVs are only H5 or H7 subtypes, while LPAIV can be any H1–H18 subtype.

Clinical Signs:

- Low Pathogenic Avian Influenza (LPAI): Symptoms are often mild or inapparent but can include respiratory issues and a 20-40% drop in egg production. Mortality rates vary with co-infections and environmental stressors.
- Highly Pathogenic Avian Influenza (HPAI): Causes sudden, high mortality (up to 100%), rapid spread, and severe clinical signs such as cessation of egg laying, respiratory distress, swelling of the head and face, hemorrhages, cyanosis and occasionally neurological symptoms.

Diagnosis:

- Virus Isolation: Inoculation of clinical samples into chick embryos.
- Serological Tests: Methods include Agar Gel Precipitation, Complement Fixation, Fluorescent Antibody Technique and ELISA.
- Molecular Techniques: PCR and RT-PCR on oropharyngeal swabs provide rapid diagnosis.

Infectious bursal disease (IBD) or Gumboro disease

Infectious Bursal Disease (IBD), also known as Gumboro disease, is a highly contagious and immunosuppressive disease affecting chickens, particularly those aged 3 to 6 weeks. It causes significant economic losses worldwide.

Aetiology:

- IBD is caused by the Infectious Bursal Disease Virus (IBDV), an RNA virus in the genus Avibirnavirus, family Birnaviridae. The virus has a non-enveloped capsid and a double-stranded RNA genome

Clinical Onset and Symptoms:

- Excretion: Virus excretion begins within 24 hours post-infection.
- Progression: Symptoms emerge 2-3 days after exposure to very virulent strains (vvIBDV), with mortality peaking and declining within 5-7 days.
- Symptoms: Severe depression, ruffled feathers, trembling, white watery diarrhea, severe prostration, vent picking, soiled vent feathers with urates, anorexia, dehydration and increased water consumption.

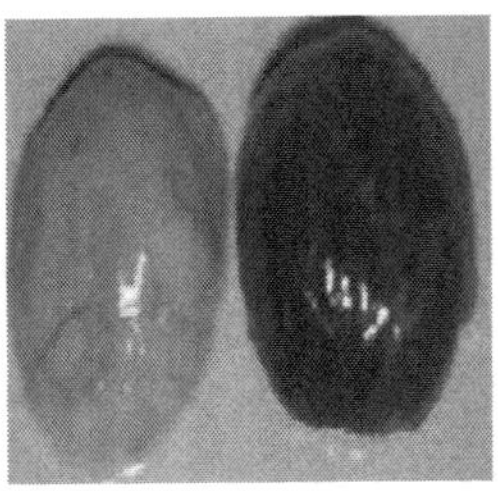

A-2 Affected bursa
Hemorrhagic to sever hemorrhagic

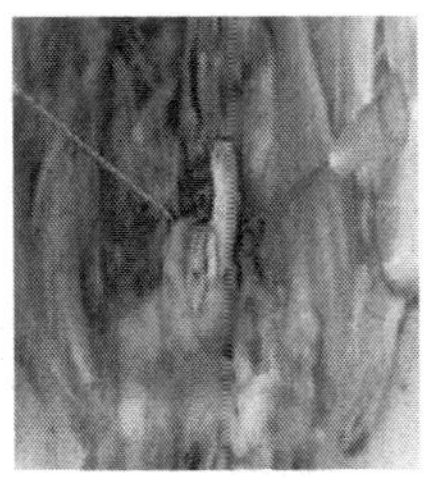

B-Caseous exduate of
Bursa of fabricious

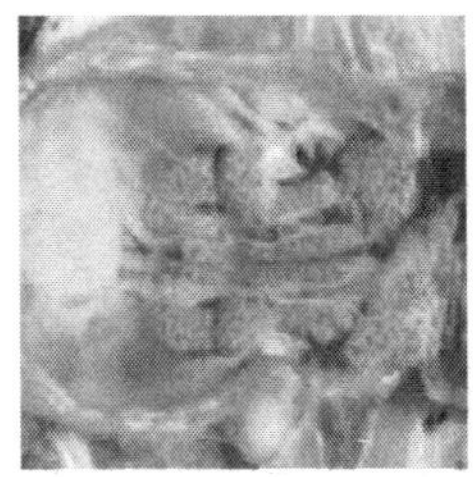

C-severe urate diathesis
Serous lesion of kidney

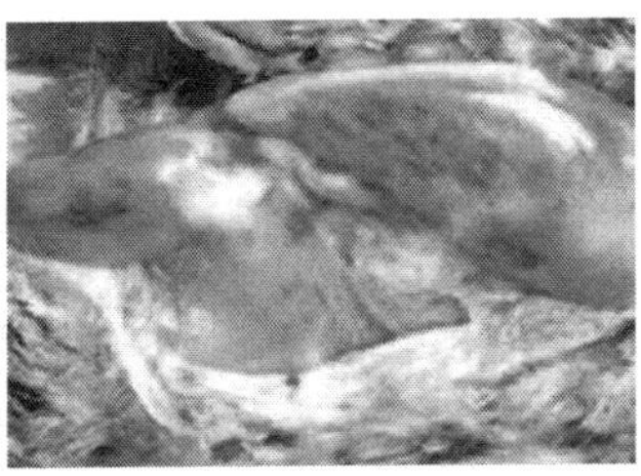

D-Hemorrhage in the affected pectoral and thigh muscle.

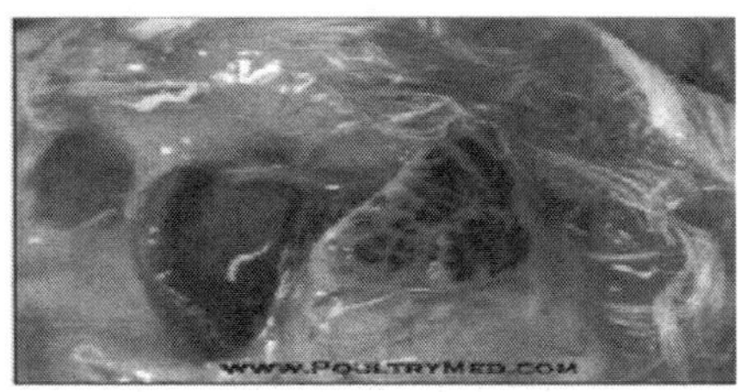

E-Liver appears pale, bile staine.

Age-Related Susceptibility:

Chickens aged 3-6 weeks are most severely affected. Birds younger than 2 weeks or older than 6 weeks show milder symptoms.

Diagnosis:

Clinical Diagnosis: Based on age, symptoms, and swollen, oedematous bursa at necropsy.

Laboratory Tests:

- Virus Isolation: Bursa of Fabricius, embryonated chicken eggs.
- Immunofluorescence: Detects viral antigens in bursa samples.

- Electron Microscopy: Visualizes the virus.
- Agar Gel Precipitation Test (AGPT) & ELISA: Detect viral antigens.
- Virus Neutralization Test: Quantifies IBDV antibodies.
- Reverse Transcriptase PCR: Targets VP1 or VP2 genes for strain characterization.
- Increased lesion incidence in MDV-infected chickens, especially on cholesterol-rich diets.

Marek's disease (MD)

Marek's Disease (MD) is a significant economic concern in poultry, characterized by rapid development of lymphoid tumors, eye conditions like *"pearl eye,"* paralysis, and immunosuppression

Aetiology:

MD is caused by the Marek's Disease Virus (MDV), classified as Gallid alphaherpesvirus 2 (GaHV-2), a member of the genus Mardivirus in the subfamily Alphaherpesvirinae, family Herpesviridae.

Table 1. Clinical Signs of Mareks disease:

Classical MD	Acute MD	Transient Paralysis	Atherosclerosis
*Paralysis from nerve enlargement due to lymphoid infiltration. ***Symptoms:** Wing and leg paralysis, torticollis, respiratory and gastrointestinal issues.	*Higher mortality (10-30% in unvaccinated flocks; up to 70% in severe outbreaks). *Sudden death or symptoms similar to classical MD.	*Sudden neck and leg paralysis, lasting around 3 days before recovery, affecting up to 50% of a flock. *Encephalitic lesions with lymphocyte infiltration and antigen-antibody complex deposition.	*Increased lesion incidence in MDV-infected chickens, especially on cholesterol-rich diets.

Diagnosis:

Based on clinical signs (paralysis, elevated mortality, tumors, enlarged nerves).

- Laboratory Tests:
- Immunohistochemistry: Identifies viral or tumor-specific antigens in tissues.
- PCR-Based Tests: Detects and quantifies the virus in clinical samples.
- Advanced PCR Methods: Utilizes nucleotide sequences from various MDV strains for precise detection and quantification.

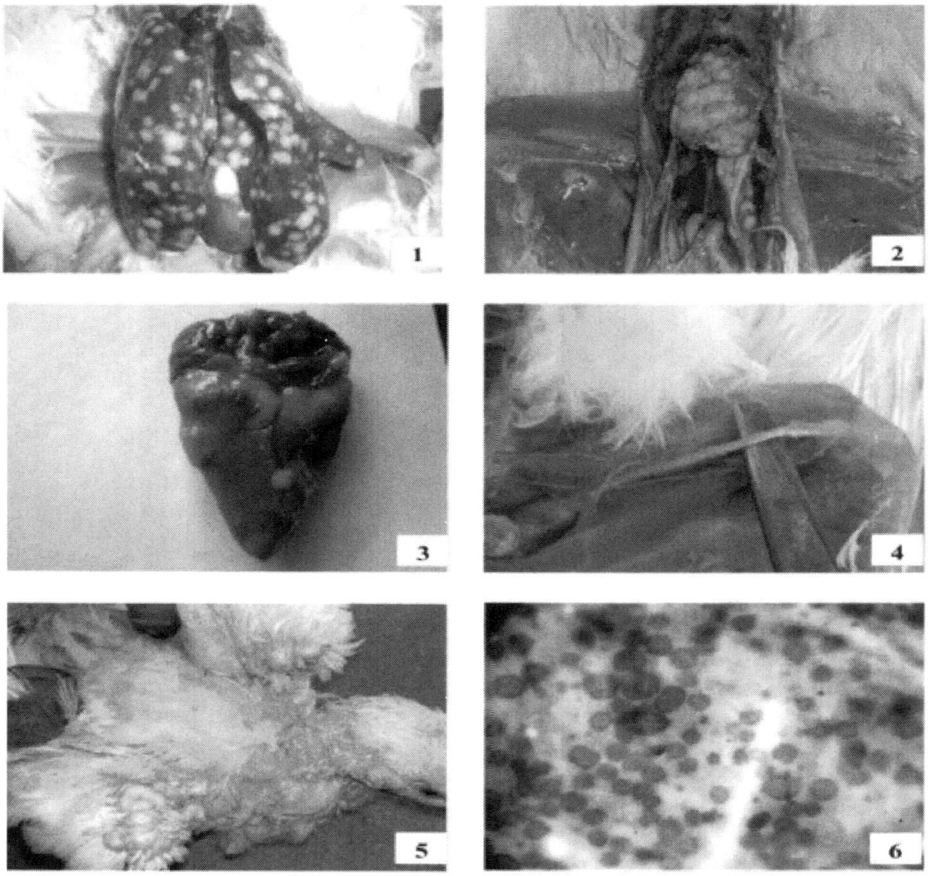

Fig 1: Liver showing multiple white nodular growths; **Fig 2**: Grayish white multiple nodular growths in ovaries with cauliflower like appearance; **Fig 3**: Heart with nodular growths; **Fig 4**: enlarged & edematous nerve **Fig 5**: Skin showing nodular lesions; **Fig 6**: Cytological smear showing pleomorphic neoplastic lymphoid cells. Giemsa stain x1000.

Infectious bronchitis (IB)

Aetiology: Infectious Bronchitis (IB) is a highly contagious poultry disease caused by the Infectious Bronchitis Virus (IBV), a member of the Gamma-coronavirus genus in the Coronaviridae family. It is characterized by respiratory distress, depression, and disheveled feathers.

Table 2. Clinical signs of Infectious Bronchitis:

Respiratory System:	Digestive and Excretory Systems:	Reproductive System:	General Symptoms:
• Chicks show respiratory distress, nasal discharge, watery eyes, swollen sinuses, and reduced feed intake. • Older chickens have milder symptoms. Broilers with nephropathogenic strains show depression, ruffled feathers, wet droppings and increased mortality. • Laying flocks may have mild respiratory signs and reduced production.	• Broilers with nephropathogenic strains show wet droppings and increased water intake. • Laying flocks experience reduced egg production, poor egg quality and lower hatchability.	• Laying flocks may see up to 70% egg production decline, eggshell abnormalities and "false layers." • The D388 strain leads to severe production drops, nephritis, cystic oviducts and up to 30% mortality in young birds.	• Depression, ruffled feathers, decreased appetite and general malaise. • Severity varies with environmental factors, bird strain, vaccination and co-infections.

Diagnosis:

- Virus Isolation: Inoculation of suspected samples into 9-10 day-old embryonated SPF eggs show stunting and embryo curling.
- Serological Tests: Virus Neutralization, Agar Gel Immunodiffusion, ELISA.
- Molecular Diagnosis: Real-time RT-PCR for virus detection and typing.

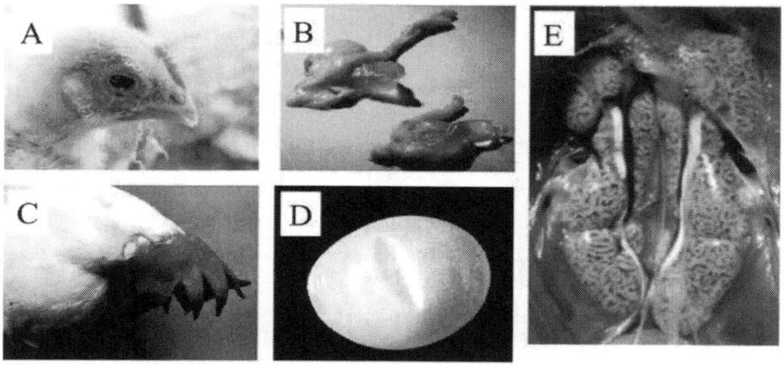

IB clinical signs and pathology. (A) Ephiphora, (B) cystic dilatation of the oviduct (C) depression, (D) thin shelled eggs and (E) enlarged kidneys

Avian Infectious Laryngotracheitis (ILT)

Aetiology: ILT is an acute, highly contagious viral disease in chickens, characterized by inflammation and hemorrhage in the larynx and trachea. It is caused by the Infectious Laryngotracheitis Virus (ILTV), which belongs to the genus Iltovirus in the subfamily Alphaherpesvirinae of the Herpesviridae family.

Clinical Signs of ILT:

- Severe Form:
  - » Dyspnea (difficulty breathing)
  - » Bloody mucus in the respiratory tract
  - » High morbidity and mortality (up to 70%)
- Mild Form:
  - » Depression
  - » Reduced egg production and weight gain
  - » Conjunctivitis
  - » Swelling of infraorbital sinuses ("almond-shaped eyes")
  - » Nasal discharge

Gross Lesions:

ILT results in inflammation, necrosis, and hemorrhage in the respiratory tract mucosa, especially in the larynx and trachea. Infected cells often contain intranuclear inclusion bodies and can form multinucleated cells (syncytia). Detached necrotic cells lead to bloody mucus in the trachea.

Diagnosis:

- Virus Isolation: Inoculation of embryonated chicken eggs via the chorioallantoic membrane (CAM) route.
- Rapid Diagnosis:
- Electron Microscopy: Identifies virus particles in tracheal samples.
- Immunofluorescence Detects viral antigens in smears or frozen tissue sections.
- ELISA or AGID Detects viral antigens in tracheal samples.
- PCR: Differentiates field strains from vaccinal strains.
- Serological Tests: Virus Neutralization Test, ELISA (Flockscreen), or AGID for antibody detection.

Inclusion Body Hepatitis (IBH)

- Affected Population: Young broilers
- Economic Impact: Substantial financial losses for farmers

Aetiology: HPS, also known as hydropericardium-hepatitis syndrome or *"Litchi heart disease,"* primarily affects young broilers and causes substantial economic losses for poultry farmers. The disease is caused by Fowl Adenovirus (FAdV) serotype 4, a Group 1 FAdV of the Adenoviridae family.

Clinical Signs:

- Non-specific symptoms like reduced growth rates and sudden increase in bird deaths.
- Specific signs include fluid accumulation in the pericardial sac and hepatitis
- Fluid accumulation in the pericardial sac (hydropericardium)
- Swollen liver with petechial hemorrhages
- Pulmonary edema
- Pale, swollen kidneys
- Enlarged spleen with hemorrhagic lesions
- Swollen bursa with hemorrhagic lesions

Diagnosis:

1. Macroscopic Examination:
- Clear fluid accumulation in the pericardium.
- Yellow subcutaneous fat and mottled, swollen liver.
- Pale, swollen kidneys and enlarged spleen.
2. Serological Tests:
- ELISA for FAdV antibodies.
- Additional tests: Agar Gel Immunodiffusion, Counter Immunoelectrophoresis, Fluorescent Antibody Techniques, Immunoperoxidase Assays, Modified ELISA.
3. Virus Isolation:
- In Vivo: Specific pathogen-free (SPF) birds.
- In Ovo: SPF embryonated eggs.
- In Vitro: Primary chicken embryo liver cells.
4. Molecular Techniques:
- Restriction Endonuclease Assay (REA).

- In Situ Hybridization with DNA probes.
- PCR targeting the hexon gene.

Fowl pox

Fowl Pox is a viral disease that affects a wide range of bird species, including commercial poultry, pet birds, and wild birds, impacting around 232 species across 23 orders. It is of economic importance due to its effects on egg production, growth, and mortality rates.

Aetiology:

- Fowl Pox is caused by Avian Pox Viruses, which belong to the genus Avipoxvirusin thePoxviridae family. The Fowl Pox Virus (FPV) is the type species of this genus.

Forms of Disease:

- Cutaneous Form: Features nodular lesions on non-feathered body parts.
- Diphtheritic Form: Presents with fibrino-necrotic and proliferative lesions in the upper respiratory tract, mouth, and esophagus.
- Systemic Form: Mainly affects canaries and is associated with high mortality rates.

Clinical Signs:

- Cutaneous Form: Nodular lesions on the comb, wattle, eyelids, and other non-feathered areas, potentially hindering feeding and drinking.
- Diphtheritic (Wet Pox) Form: Yellowish lesions (cankers) on mucous membranes of the mouth, esophagus, or trachea.
- Respiratory Symptoms: Similar to those caused by infectious laryngotracheitis virus, with lesions in the mouth and trachea affecting eating, drinking and breathing.

Diagnosis:

- Histopathology: Examination of tissue sections from lesions to identify cytoplasmic inclusions.
- Electron Microscopy: Visualization of virus particles in lesions and exudate.

Isolation and Identification

- Bird Inoculation: Induces cutaneous lesions in susceptible birds within 5–7 days.
- Chicken Embryo Inoculation: Examines for pock lesions on the chorioallantoic membrane (CAM) after 5–7 days.

- Cell Culture: Virus isolation through primary or secondary cell cultures of avian origin.
- Molecular Methods: PCR, RFLP and genomic probes

Chicken Infectious Anemia

Aetiology: Chicken Infectious Anemia (CIA) was first reported in 1979 in Japan and is caused by a small DNA virus belonging to the genus Gyrovirus in the family Circoviridae. The primary hosts are young chicks, but antibodies have been found in quails, and the virus has been isolated from turkeys.

Clinical signs:

- Affects chickens of all ages but is most severe in day-old chicks.
- Symptoms include poor weight gain, severe anemia, bone marrow aplasia, and lymphoid atrophy.
- Immunosuppression: Leads to increased susceptibility to secondary infections, reduced vaccine efficacy, and compromised production performance. Mortality is typically 5-10% but can reach up to 60% with secondary complications.

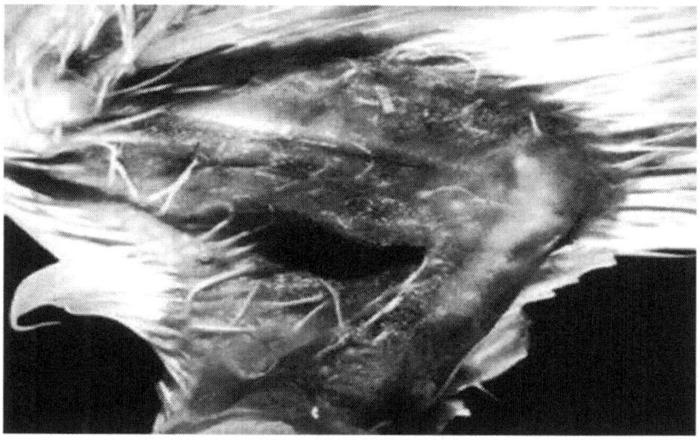

Fig: Lesions in chickens associated with chicken infectious anemia and hemorrhagic anemia disease. Gangrenous dermatitis (blue wing disease)

Immunosuppression:

- CIA leads to generalized lymphoid atrophy and suppression of T-cell functions, resulting in increased susceptibility to secondary infections and vaccine failures.
- It also plays a role in multifactorial diseases like blue wing disease and gangrenous dermatitis.

Post mortem Lesions:

- Gross Lesions: Include thymus atrophy, pale/fatty bone marrow, and swollen/mottled liver.
- Microscopic Lesions: Involve lymphoid aplasia, bone marrow hypoplasia, liver necrosis, and apoptosis in various organs.

Diagnosis:

- Tentative Diagnosis: Based on flock history, clinical signs, hematological changes, and gross pathological findings.
- Confirmatory Diagnosis: Requires isolation and identification of the virus.
- Seromonitoring: Methods include ELISA, virus neutralization (VN) and immunofluorescent assay (IFA).
- Advanced Techniques: PCR, restriction enzyme (RE) analysis and sequencing for genomic characterization.

Avian Encephalomyelitis

Aetiology: Avian Encephalomyelitis Virus (AEV): A picornavirus primarily targeting the central nervous system (CNS) of chickens, transmitted through the oral-fecal route. It spreads both vertically (from parent to offspring) and horizontally (from bird to bird) and remains infectious in contaminated environments for extended periods.

Pathogenesis:

- Causes neurological symptoms in chickens, including paralysis, ataxia, and muscular dystrophy, especially in young birds.
- Leads to CNS lesions characterized by neuron degeneration and lymphoid cell infiltration.
- Can be cultured in chicken embryos, producing characteristic symptoms and lesions.

Clinical Signs:

- Young Birds (Under 3–5 Weeks): Show neurologic symptoms such as head and neck tremors, ataxia, and possible paralysis. Severe cases may result in starvation and death; survivors may develop cataracts by 18-20 weeks.
- Adult Birds: Generally asymptomatic but may experience a temporary decline in egg production and hatchability

Control and Diagnosis:

- Diagnosis has improved with the development of enzyme-linked immunosorbent assays (ELISA) for detecting antibodies.
- Vaccination with live, highly enterotropic AEV strains has reduced disease incidence.

Avian Leucosis

Aetiology

- Virus Classification: Alpharetrovirus genus, Retroviridae family: Possess reverse transcriptase for forming DNA provirus integrated into host genome.
- Genome: It includes gag, pro, and env structural genes, along with regulatory sequences, all flanked by long terminal repeats (LTRs).

Table 3. Forms of Avian leucosis

Lymphoid leukosis	Erythroid leukosis	Myeloid Leukosis	Myeloblastic Form	Myelocyctic form
Enlargement of liver (with miliary, diffuse, or nodular foci), bursa of Fabricius, spleen and other organs *Microscopic Pathology: Clusters of immature lymphoid cells (lymphoblasts) coalesce, with B-cells expressing IgM.	Moderately enlarged liver and spleen, bright cherry-red color, bright red liquid bone marrow, anemia, muscle hemorrhages, abdominal hemorrhage *Microscopic Pathology: Intrasinusoidal accumulations of erythroblasts in liver, spleen, and bone marrow.	*Forms: Myeloblastic (myeloblastosis) and myelocytic (myelocytomatosis). *Prevalence: Particularly prevalent in broiler breeders since 1996 due to ALV subgroup J.	Enlarged liver and spleen, yellowish-grey granular appearance of liver, firm reddish-grey bone marrow. *Microscopic Pathology: Intravascular and extravascular accumulations of immature myeloid cells.	Yellowish-white myelocytomas on skeleton and internal organs. *Microscopic Pathology: Well-differentiated myelocytes with cytoplasmic granules.

Detection and Diagnosis

- Biological and molecular assays are used for detecting and classifying ALV.
- Assays for ALV include detection of specific proteins (p27) or viral RNA/DNA using ELISA and PCR.
- Differentiating between endogenous and exogenous ALV is challenging due to shared antigens.

Avian Viral Arthritis

- Avian Reoviruses: Common in poultry, often asymptomatic but can cause diseases like viral arthritis/tenosynovitis, primarily in chickens.

Overview:

- Viral Arthritis/Tenosynovitis: Affects meat-type chickens, causing lameness. Other associated conditions include enteric problems, respiratory disease and malabsorption syndrome.

Aetiology: Avian Reoviruses: Belong to the Orthoreovirus genus with double-stranded RNA genomes consisting of ten segments. Strains vary in pathogenicity and antigenic types.

Clinical Signs: Viral Arthritis/Tenosynovitis: Main clinical manifestation in broiler chickens, characterized by swelling of hock joints, acute lameness and tendon rupture.

- Other Conditions: Include enteric problems and respiratory diseases.

Diagnosis:

- Virus Isolation: Confirms the presence of reovirus in clinical samples but does not necessarily indicate causation.
- Serology: Techniques like ELISA detect antibodies but are complicated by widespread infection.
- Molecular Approaches: PCR and PCR-RFLP identify viral genetic material, offering rapid and sensitive diagnostics.
- Histopathology: Reveals characteristic joint tissue changes due to reovirus infection.

Prevention and Control:

- Vaccination: Breeder vaccination with inactivated vaccines provides maternal antibodies to chicks.
- Biosecurity and Management: Key measures to control reovirus-associated diseases.

Swollen Head Syndrome

- Swollen Head Syndrome (SHS): A form of colibacillosis in poultry, characterized by acute to subacute cellulitis affecting the periorbital and surrounding subcutaneous tissues of the head.

Aetiology:

- Initially linked to *E. coli* strains from the air sac, trachea, and infraorbital sinus. Later found to involve a pneumovirus causing acute rhinitis, followed by *E. coli* invasion into facial subcutaneous tissues.

Clinical signs:

- Early Symptoms: Snicking, conjunctivitis, swelling around the eyes, lacrimal gland and subcutaneous edema of the head. Additional signs include swelling of the peri- and infraorbital sinuses, torticollis and neurological disorders.

- Progression: Symptoms appear within 4-5 days, with a 2-3% reduction in egg production and 3-4% mortality over 2-3 weeks. Mortality rates range from 3-7%.

Diagnosis and Identification:

- Streptococcus spp.: Identified through phenotypic methods and 16S rRNA sequencing.
- Rapid Detection: Universal primer multiplex PCR (UP-MPCR) used for detecting *P. aeruginosa.*

Infectious stunting syndrome

- Infectious Stunting Syndrome (ISS): First reported in England in 1981, similar to a syndrome previously described in the Netherlands. It causes significant financial losses in broilers, turkeys and certain layer breeds. ISS leads to uneven growth, partial recovery and mortality between 6 and 14 days old.

Aetiology:

- The exact causative agent is unidentified; reoviruses and caliciviruses have been isolated but are not solely responsible. The involvement of fish meal and mycotoxins has been investigated.

Clinical signs:

- Broilers: Stunted growth with abdominal distension, well-fleshed for their size, uneven growth and delayed plumage development.
- Turkey: Abnormal feathering ("helicopter" feathers), skeletal abnormalities (enlarged rib heads, thickened growth plates) and pendulous abdomens due to distended intestines. Partial recovery after 5 weeks of age.

Diagnosis:

- Based on clinical signs, transmission studies, and pathological findings. Histopathology reveals lesions in the pancreas, bursa of Fabricius, thymus, proventriculus and intestines. Experimental infections with reoviruses are inconclusive, suggesting other agents may be involved.

Avian Monocytosis

Aetiology: Avian Monocytosis: A sudden-onset condition in chickens early in their laying period, characterized by high morbidity and variable mortality (0-50%). The disease is marked by an increase in circulating monocytes and was more common before 1960.

Clinical Signs:

- Depression, loss of appetite, distended crop, darkened comb and wattles, watery diarrhea, dehydration and reduced egg production.

Diagnosis:

- Based on history, clinical signs and post-mortem findings while excluding other conditions such as fowl cholera, vibriosis, capillariasis, coccidiosis, infectious bursal disease (IBD) and typhoid.

Points to be remembered

- Newcastle Disease (ND): Affects multiple systems (respiratory, gastrointestinal, and nervous), leading to high mortality and reduced egg production.
  - Velogenic Strains: Highly fatal, cause severe symptoms.
  - Mesogenic Strains: Moderate virulence with low mortality.
  - Lentogenic Strains: Mild symptoms, mainly respiratory.
- Asymptomatic Enteric Isolates: Subclinical infections.
- Avian Influenza (AI): Highly contagious, causing severe respiratory symptoms and high mortality. It is Caused by type A influenza viruses (*Orthomyxoviridae* family).
- Marek's Disease (MD): Causes tumors, paralysis, and immunosuppression, particularly affecting young chickens. It is caused by Marek's Disease Virus (*Herpesviridae*). Paralysis, tumors and eye conditions ("pearl eye").
- Infectious Bronchitis (IB): Primarily affects the respiratory system and causes decreased egg production. Respiratory issues, watery eyes, nasal discharge and reduced egg production.
- Immunosuppressive diseases like MD and IBD lead to secondary infections, further increasing losses.
- Diagnosis is mainly by Virus isolation and molecular methods (RT-PCR, ELISA) are common for detecting AI, ND, MD and IB.
- Key control measures include vaccination, strict biosecurity practices (hygiene, controlling bird movement) and proper disposal of dead birds.
- Vaccination is essential for controlling ND, IB, and MD, but challenges like viral mutations persist.
- Surveillance and early detection are vital for managing outbreaks and reducing disease spread.

References

Ahmed, R., Deka, P., Hazarika, R., Barua, J., Sharma, A., Sarma, J., Devi, B., Das, S., Nath, M.K., Das, G. and Sarma, M., 2022. Viral diseases of poultry in Assam, India: A review. *IJBSM.*, 13: 943-953.

Alexander, D.J., 2008. Avian influenza–diagnosis. *Zoonoses and public health*, 55: pp.16-23.

Bracewell, C.D. and Randall, C.J., 1984. The infectious stunting syndrome. *J. World's Poult. Sci.*, 40: 31-37.

Cattoli, G., Susta, L., Terregino, C. and Brown, C., 2011. Newcastle disease: a review of field recognition and current methods of laboratory detection. *J. Vet. Diagn. Invest.*, 23:637-656.

Chubb, L.G. and Gordon, R.F., 1957. The avian leucosis complex-a review.

Das, S., Das, D., Panda, S.K., Sagarika, S. and Jena, B., 2018. Clinico-pathological studies of marek's disease in chickens. *Int. J. Livest. Res*, 8:207-217.

Ficken, M.D., Nasisse, M.P., Boggan, G.D., Guy, J.S., Wages, D.P., Witter, R.L., Rosenberger, J.K. and Nordgren, R.M., 1991. Marek's disease virus isolates with unusual tropism and virulence for ocular tissues: clinical findings, challenge studies and pathological features. *Avian Pathol.*, 20: 461-474.

Getabalew, M., Alemneh, T., Akeberegn, D., Getahun, D. and Zewdie, D., 2019. epidemiology, Diagnosis & Prevention of Newcastle disease in poultry. *Am J Biomed Sci Res.*, 16: 50-9.

Jackwood, M.W. and de Wit, S., 2013. Infectious bronchitis. *Poult Sci.*, 139-159.

Oluwayelu, D.O., 2010. Diagnosis and epidemiology of chicken infectious anemia in Africa. *Afr. J. Biotechnol., 9*: 2043-2049.

Payne, L.N. and Venugopal, K., 2000. Neoplastic diseases: Marek's disease, avian leukosis and reticuloendotheliosis. *Rev. sci. tech.*, 19: 544-560

Powell, P.C., 1986. Marek's Disease–a world poultry problem. *J. World's Poult. Sci.*, 42205-218.

Radwan, A.E.H., Fathy Mohamed, M. and Kamal Eldin Ahmed, A., 2018. Bacteriological studies on bacterial pathogens isolated from broiler chickens with swollen head syndrome. *J. Vet. Med. Res.*, 25: 191-198.

Schat, K.A. and van Santen, V.L., 2020. Chicken infectious anemia and circovirus infections in commercial flocks.*Poult. Sci*9: 284-320.

Tannock, G.A. and Shafren, D.R., 1994. Avian encephalomyelitis: A review. *Avian Pathol., 23*: 603-620.

Tripathy, D.N. and Reed, W.M., 2013. Pox. *Poult Sci.*, 333-349.

Van der Heide, L., 1977. Viral arthritis/tenosynovitis: a review. *Avian pathol.*, 6: 271-284.

Wagari, A., 2021. A review on infectious bursal disease in poultry. *Health Econ.*, 7: 167.

Zhou, W., Shen, B., Yang, B., Han, S., Wei, L., Xiao, B. and Zhou, J., 1997. Isolation and identification of chicken infectious anemia virus in China. *Avian Dis.*, 361-364.

CHAPTER 9

PROTOZOAL AND PARASITIC DISEASES OF POULTRY

[1*]Uppari Anusha Rao and [2]M Bindu Pushpa

[1] Assistant Professor, Department of Veterinary Parasitology, Institute of Veterinary Science & Animal Husbandry, SOA DU, Bhubaneswar, Odisha

[2] Assistant Professor, Department of Veterinary Public Health, Institute of Veterinary Science & Animal Husbandry, SOA DU, Bhubaneswar, Odisha

**Corresponding author- uppari.anusha97@gmail.com*

Abstract

Poultry parasitic illnesses provide serious obstacles to the well-being, production, and health of the birds. Numerous parasites, such as helminths, ectoparasites, and protozoa, are responsible for these illnesses. The most common is coccidiosis, which is brought on by several Eimeria species and causes severe intestinal damage as well as significant mortality rates, particularly in young birds. Ascaridiasis and Capillariasis are examples of helminthic illnesses that result in poor nutrient absorption, weight loss, and decreased egg production. Lice, mites, and fleas are examples of ectoparasites that cause discomfort, stress, and anaemia, which further impairs the health and productivity of birds. Adhering to stringent cleanliness protocols, implementing biosecurity protocols, doing routine health assessments, and utilizing antiparasitic drugs and immunizations are all beneficial control and preventative tactics. These coordinated strategies are necessary to reduce the negative effects of parasite infections on chickens.

Keywords: Poultry, Parasitic Diseases, Coccidiosis, Helminths, Ectoparasites

Introduction

Parasitic diseases are a major concern in poultry farming, significantly affecting bird health, productivity and overall farm economics. Numerous parasites, such as helminths (worms), ectoparasites (external parasites), and protozoa, are responsible for these illnesses., each presenting unique challenges and requiring specific management strategies. Understanding and controlling these diseases is crucial for maintaining healthy flocks and ensuring sustainable poultry production. The main objective is to understand the different types of parasites affecting poultry, and to explore how these diseases affect bird health, growth, productivity and mortality rates. As this is a book on guide to poultry science, the chapter on poultry parasitic disease is commited to discussion of important parasitic diseases to learn about the causes, clinical signs, symptoms and diagnostic methods for identifying parasitic infections in poultry their prevention and control measures.

Protozoan Diseases

Coccidiosis

Protozoa belonging to the genus *Eimeria* are the source of coccidiosis, a parasitic disease that affects poultry's digestive tract. Because of its effects on both bird health and farm economics, it is one of the most important illnesses in poultry production.

Genus and Species: The disease is caused by several *Eimeria* species. Each species targets specific parts of the intestine:

Eimeria tenella: Caecum

Eimeria necatrix: Small intestine and caecum

Eimeria maxima: Mid-intestine

Eimeria acervulina: Upper part of the small intestine

Eimeria brunetti: Lower small intestine, rectum, cloaca

Eimeria mitis and *Eimeria praecox*: Small intestine, less pathogenic

Eimeria mivati: Mostly upper small intestine, but can range from duodenum through most of intestinal tract

Life cycle: The lifecycle starts with the ingestion of sporulated oocysts by the bird.Oocysts require specific environmental conditions to sporulate and become infective, including oxygen, moisture and a suitable temperature (around 27-30°C). Inside the bird, the oocysts release sporozoites that invade the intestinal cells. These undergo several asexual replication cycles (schizogony) and then sexual reproduction (gametogony), forming new oocysts that are shed in the feces.

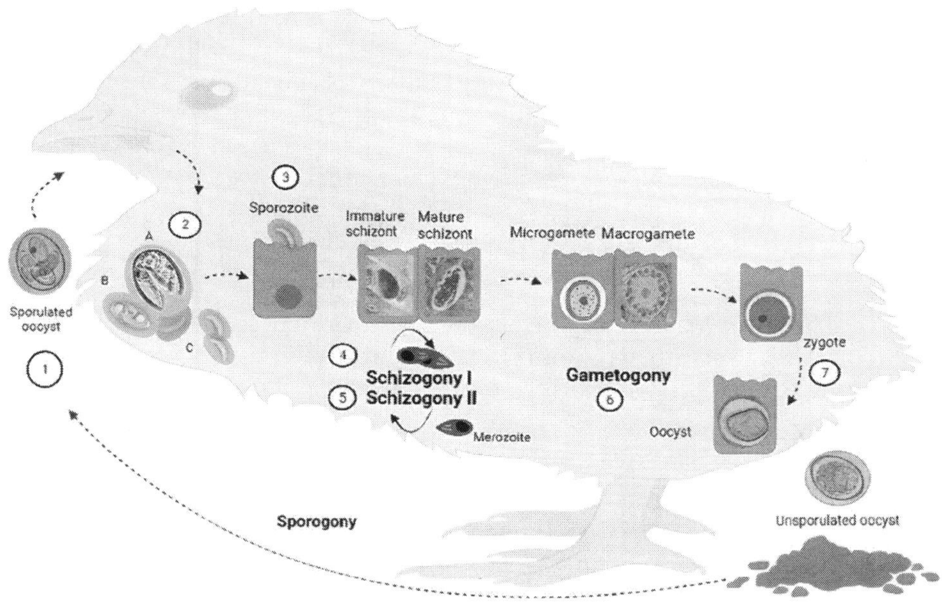

Fig 1. Life cycle of *Eimeria spp*. **in chickens** (*source-ndvsu.org/images)*

Transmission: Birds contract the infection by consuming sporulated oocysts from tainted feed, water, litter, tools, or surroundings.Oocysts can last for extended periods of time due to their great resistance to environmental factors and disinfectants.

Symptoms: Intestinal Symptoms like Diarrhea, often bloody or mucoid, especially with *E. tenella* and *E. necatrix*. Dehydration due to fluid loss. Weight loss and poor growth. Behavioral change includes lethargy and decreased activity. Ruffled feathers and huddling behavior. Reduced feed intake and reluctance to move.A discerible decrease in egg output in laying hens and poor eggshell quality.

Mortality: Acute Infections can lead to sudden death, particularly in young birds and chronic Infections often result in stunted growth and poor overall performance.

Diagnosis: Identifying characteristic symptoms and considering flock history, including previous outbreaks and environmental conditions.

Post-Mortem Examination: Examination of the intestines for typical lesions associated with different *Eimeria* species.

E. tenella: Hemorrhagic lesions in the caecum.

E. necatrix: White and red spots in the mid-intestine, ballooning of the intestines. (Appearce described as salt and pepper)

E. maxima: Thickening of the intestinal wall, orange mucus.

Microscopic Examination: Detection of oocysts using flotation techniques. Intestinal scrapings for Identification of various lifecycle stages under a microscope.

Treatment and Control: Anticoccidial Drugs- Coccidiostats drugs that inhibit the growth of *Eimeria* but do not kill them, allowing the birds to develop immunity (e.g., amprolium, sulfaquinoxaline).Coccidiocidal drugs that kill the parasites (e.g., toltrazuril, diclazuril). These drugs are typically added to feed or water.

Vaccination by live attenuated vaccines which contain weakened strains of *Eimeria* that stimulate immunity without causing severe disease.Vaccines usally administered to chicks at a young age to build immunity.

Regular cleaning and replacing of litter to reduce oocyst load.Ensuring good air circulation to keep the environment dry. Limiting access to poultry houses, disinfecting equipment and controlling the movement of people and animals.

Probiotics and prebiotics can enhance the gut microflora, improving resistance to infections. Natural additives like essential oils, herbal extracts, and other natural compounds may have anticoccidial effects.

Prevention: Good Farm Management Practices, sanitation by regular cleaning and disinfection of poultry houses. Avoiding overcrowding to reduce stress and the spread of disease. Providing a well-balanced diet to support overall health and immunity.Rotation of Anticoccidial drugs. Rotating between different classes of anticoccidial drugs to prevent the emergence of *Eimeria* strains resistant to drugs.

Table 1. Characteristics of Chicken Coccidia

DIFFERENTIAL CHARACTERISTICS FOR 9 SPECIES OF CHICKEN COCCIDIA ◉ — DIAGNOSTIC CHARACTERISTICS IN RED — SPECIES OF DOUBTFUL VALIDITY

	CHARACTERISTICS	*E. acervulina*	*E. brunetti*	*E. maxima*	*E. mitis* †	*E. mivati* ‡	*E. necatrix*	*E. praecox*	*E. tenella*	*E. hagani*
MACROSCOPIC LESIONS	ZONE PARASITIZED	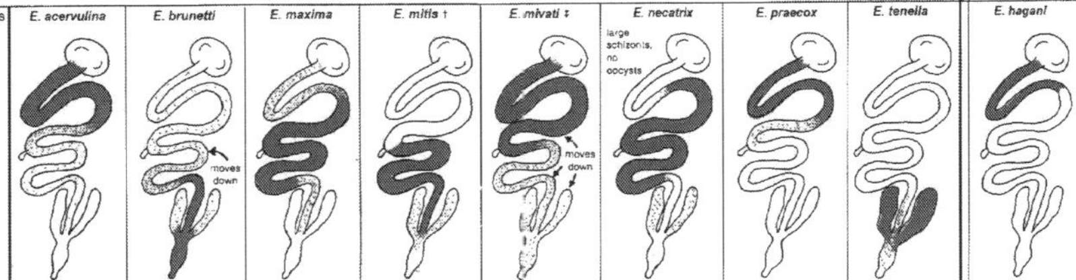	moves down			moves down	large schizonts, no oocysts			
	MACROSCOPIC LESIONS	light infection: whitish round lesions sometimes in ladder-like streaks heavy infection: plaques coalescing, thickened intestinal wall	coagulation necrosis mucoid, bloody enteritis in lower intestine	thickened walls, mucoid, blood - tinged exudate, petechiae	no discrete lesions in intestine. mucoid exudate	light infection: rounded plaques of oocysts heavy infection: thickened walls coalescing plaques	ballooning, white spots (schizonts), petechiae, mucoid blood - filled exudate	no lesions, mucoid exudate	onset: hemorrhage into lumen later: thickening, whitish mucosa, cores clotted blood	pinhead hemorrhages petechiae
MICROSCOPIC CHARACTERISTICS	MILLIMICRONS / OOCYSTS REDRAWN FROM ORIGINALS	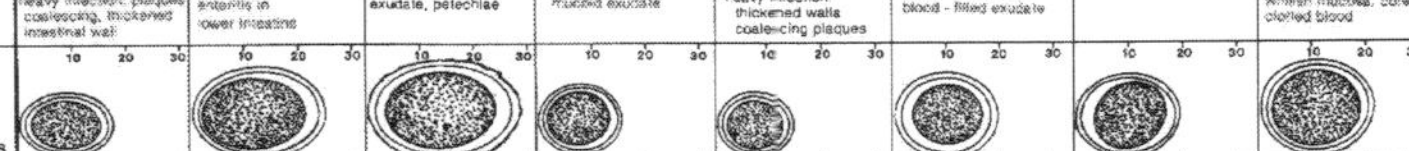								non available
	LENGTH × WIDTH μ LENGTH = WIDTH =	AV = 18.3 × 14.6 17.7 - 20.2 13.7 - 16.3	24.6 × 18.8 20.7 - 30.3 18.1 - 24.2	30.5 × 20.7 21.5 - 42.5 16.5 - 29.8	16.6 × 14.2 11.7 - 18.7 11.0 - 18.0	15.6 × 13.4 11.1 - 19.9 10.5 - 16.2	20.4 × 17.2 13.2 - 22.7 11.3 - 18.3	21.3 × 17.1 19.8 - 24.7 15.7 - 19.8	22.0 × 19.0 19.5 - 26.0 16.5 - 22.8	19.1 × 17.6 15.8 - 20.9 14.3 - 19.5
	OOCYST SHAPE AND INDEX-LENGTH/WIDTH	ovoid 1.25	ovoid 1.31	ovoid 1.47	subspherical 1.09	ellipsoid to broadly ovoid 1.16	oblong ovoid 1.19	ovoidal 1.24	ovoid 1.16	broadly ovoid 1.08
	SCHIZONT, MAX IN MICRONS	10.3	30.0	9.4	15.1	17.3	65.9	20	54.0	
	PARASITE LOCATION IN TISSUE SECTIONS	epithelial	2nd generation schizonts subepithelial	gametocytes subepithelial	epithelial	epithelial	2nd generation schizonts subepithelial	epithelial	2nd generation schizonts subepithelial	epithelial
LIFE HISTORY CHARACTERISTICS	MINIMUM PREPATENT PERIOD-HR	97	120	121	93	93	138	83	115	99
	SPORULATION TIME MINIMUM (HR)	17	18	30	15	12	18	12	18	18

† = From Norton and Joyner (1980)
‡ = As described by Edgar and Siebold (1964)
◉ = Compiled from various sources

Peter L. Long and W. Malcolm Reid
Department of Poultry Science
The University of Georgia, Athens

Table 2. Other Protozoal infection in Poultry

Protozoa Disease	Causative Agent	Mode of transmission	Symptoms	Diagnosis	Treatment
Histomoniasis (Blackhead)	*Histomonasmeleagridis*	Fecal oral, cecal worm vector	Yellow droppings, depression, reduced feed intake, mortality	Histopathology, PCR	Antihistomonaldrugs (e.g., Metronidazole)
Trichomoniasis (Canker)	*Trichomonas gallinae*	Contaminated water/ feed	White or yellowish plaques in the mouth, difficulty in eating, weight loss	Wet mount microscopy, PCR	Antiprotozoal drugs (e.g., Metronidazole) Good hygiene, avoid contaminated water
Toxoplasmosis	*Toxoplasma gondii*	Ingestion of oocysts	General weakness, respiratory distress, mortality	Serology, PCR	No specific treatment; supportive care
Cryptosporidiosis	*Cryptosporidium* spp.	Fecal oral	Watery diarrhea weight loss, dehydration	Fecal examination, PCR	No specific treatment; supportive care
Trypanosomosis	*Trypanosome* spp	Insect vectors	Anaemia, Emaciation	Blood smaer, PCR	Antiprotozoals (diminazene)
Avian Malaria	*Plasmodium* spp.	Mosquito vectors	Rise in temperature, greenish diarrhoea, gradual emaciation, anaemia, paralysis.	Blood smear, PCR	Antimalarials (chloroquine)
Sarcocystosis	*Sarcocystis*spp.	Ingestion of sporocysts in contaminated food, water	No characteristic signs, lameness	Histopathology, serology	No specific treatment, improved sanitation

Endoparasitic infections

A. **Nematodes of poultry**

B. **Cestodes of poultry**

C. **Trematodes of poultry**

A. Nematodes of Poultry

Nematodes, or roundworms, are a significant concern in poultry farming, causing various health issues and economic losses. Here's a detailed overview of common nematodes affecting poultry, their lifecycle, clinical signs, diagnosis, and treatment.

Table 3. Common Nematodes of Poultry

Parasite	Bird Host	Lifecycle	Organs infected	Pathogenecity
Ascaridia galli (Large Roundworm)	Chicken, turkey, duck, quail	Direct	Small intestine	Moderate
Ascaridiadis similis	Turkey	Direct	Small intestine	Moderate
Amidostomum anseris	Duck, goose, pigeon	Direct	Gizzard	Severe
Capillaria caudinflata (columbae)	Chicken, turkey, duck, game birds, pigeon	Earthworms	Small intestine	Moderate to severe
Capillaria contorta (annulata)	Chicken, turkey, duck, game birds	Direct or earthworms	Mouth, oesophagus, crop	Severe
Capillaria obsignata	Chicken, turkey, goose, pigeon, quail	Direct	Small intestine, caecum	Severe
Tetrameres americana	Chicken, turkey, duck, game birds, pigeon	Grasshoppers, cockroaches	Proventriculus	Moderate to severe
Dispharynx nasuta	Chicken, turkey, game birds, pigeon	Sowbugs	Proventriculus	Moderate to severe
Ornithostrongylus quadriradiatus	Pigeon, dove	Direct	Small intestine	Severe
Gongylonema ingluvicola	Chicken, game birds	Beetles and cockroaches	Crop, oesophagus, proventriculus	Mild
Heterakis gallinarum (Caecal worm)	Chicken, turkey, duck, game birds	Direct	Caecum	Mild but transmits histomoniasis

Parasite	Bird Host	Lifecycle	Organs infected	Pathogenecity
Cyrnea colini	Turkey, game birds	Cockroaches	Proventriculus	Mild
Subulura brumpti	Chicken, turkey, duck, game birds	Earwigs, grasshoppers, beetles, cockroaches	Caecum	Mild
Cheilospirura hamulosa	Chicken, turkey, game birds	Grasshoppers, beetles	Gizzard	Moderate
Trichostrongylus tenius	Chicken, turkey, duck, game birds, pigeon	Direct	Caecum	Severe

General Preventive Measures: Hygiene and sanitation by regular cleaning and disinfection of poultry houses. Proper disposal of feces and litter. Management Practices like rotational grazing to reduce environmental contamination. Avoid overcrowding to reduce stress and the spread of infections. Regular monitoring and deworming, fecal examinations. Scheduled deworming programs. Implementing strict biosecurity measures to prevent the introduction of nematode eggs or larvae. Controlling access to poultry houses and limiting contact with wild birds and other potential carriers. Reducing intermediate host populations (earthworms, insects) through environmental controls.

B. Cestodes of Poultry

Cestodes, commonly known as tapeworms, are parasitic flatworms that can infect poultry, causing a range of health issues. Here is an overview of cestodes affecting poultry, including their lifecycle, clinical signs, diagnosis and treatment:

Common Cestodes in Poultry

Raillietina spp.

Davainea proglottina

Choanotaenia infundibulum

Amoebotaenia cuneata

Lifecycle: The lifecycle of cestodes involves two hosts, an intermediate host (often an insect) and a definitive host (the poultry). Here's a general outline, Eggs from the feces of the infected bird.Intermediate host (like beetles, ants, or snails) ingests the eggs. Inside this host, the eggs hatch into larvae (oncospheres) and then develop into cysticercoids.Poultry (definitive host) infected by eating the intermediate host containing cysticercoids. Adult tapeworms develop in the intestines of the bird, where they attach to the intestinal wall and mature, starting the cycle anew.

Clinical Signs: Weight loss or poor weight gain, decreased egg production, diarrhea or abnormal droppings, Anaemia and weakness, decreased feed conversion efficiency, Intestinal blockage in severe cases, Visible segments of tapeworms in the feces

Diagnosis: Diagnosis of cestode infection in poultry involves, Microscopic examination of feces to detect tapeworm eggs or proglottids (tapeworm segments). Necropsy by examination of the intestines of deceased birds to identify adult tapeworms.

Treatment of cestode infections in poultry includes: Anthelminthics like

- Praziquantel: Effective against a wide range of cestodes.
- Benzimidazoles: Such as fenbendazole and mebendazole, can be used, although they are more commonly used against nematodes.

Management and Control: Regular cleaning and sanitation to reduce intermediate host populations. Managing the environment to prevent birds from accessing intermediate hosts. Rotation of pasture or confinement systems to break the lifecycle of the parasites.

Preventive Measures: Regular deworming programs. Monitoring fecal samples regularly to catch infections early.

C. Trematodes of Poultry

Trematodes, commonly known as flukes, are parasitic flatworms that can infect poultry, causing various health issues. Here is an overview of trematodes affecting poultry, including their lifecycle, clinical signs, diagnosis and treatment:

Common Trematodes in Poultry

Prosthogonimus spp. (e.g., *Prosthogonimus macrorchis, Prosthogonimus ovatus*)

Echinostoma spp.

Lifecycle: Trematodes have several hosts during their life cycle, including one or more intermediate hosts (often insects and snails) and a definitive host (poultry). The diseased bird excretes eggs in its feces. Usually a snail, the first intermediate host to consume the eggs is it. The eggs hatch into miracidia inside the snail and these organisms grow into sporocysts, rediae and cercariae. After being freed from the snail, cercariae might encyst on plants or in another intermediate host, like an insect or another snail, to generate metacercariae. The definitive host, poultry, contracts the infection by consuming tainted food or water containing metacercariae. Depending on the species, adult flukes develop in the bird's intestines or other organs where they mature and re-start the cycle.

Clinical Signs: Diarrhea, weight loss or poor weight gain and reduced egg production, Lethargy, weakness and anemia. In severe cases, obstruction or damage to the affected organs (e.g., bile ducts, intestines)

Diagnosis: Diagnosis of trematode infection in poultry involves, microscopic examination of feces to detect trematode eggs.

Necropsy: Identification of adult flukes via examination of the intestines or other afflicted organs in deceased birds.

Clinical Signs: Observing clinical signs and correlating them with fecal examination findings.

Treatment: Treatment of trematode infections in poultry includes, Anthelmintics like

Praziquantel: Effective against a wide range of trematodes.

Albendazole: Sometimes used for flukes, although it is more commonly used against nematodes.

Triclabendazole: Specifically effective against liver flukes, though its use in poultry is less common.

Management and Control: Regular cleaning and sanitation to reduce intermediate host populations. Managing the environment to prevent birds from accessing intermediate hosts. Avoiding grazing on wetlands or areas with high snail populations.

Preventive Measures: Regular deworming programs. Monitoring fecal samples regularly to catch infections early. Implementing biosecurity measures to prevent the introduction of new infections.

Ectoparasites of Poultry

Parasites that live on the outside of the host, and they can significantly impact poultry health and productivity. Common ectoparasites affecting poultry include mites, lice, fleas, and ticks. Here is an overview of these ectoparasites, including their effects, identification and treatment options:

Table 4. Common Ectoparasites of Poultry

Parasite	Clinical signs	Diagnosis	Treatment	Control
Lice (e.g., Wing louse-*Lipeurus caponis*, Shaft louse- Menopon gallinae, Head louse-*Culcogaster heterrographa*, Fluff louse- *Goniocotes gallinae*, Large chicken louse- Goniodes gigas)	Feather damage, irritation, reduced egg production, weight loss	Visual inspection, finding lice or eggs on feathers, especially around the vent	Permethrin dust, carbaryl powder, ivermectin	Regular inspection, good sanitation, changing bedding regularly, treating infested birds

Parasite	Clinical signs	Diagnosis	Treatment	Control
Flea (e.g., *Echidnophaga gallinacea*)	Anemia, restlessness, decreased egg production, flea dirt on skin	Visual inspection, finding fleas or flea dirt on skin, especially around the head and eyes	Insecticidal dusts or sprays, ivermectin, flea collars (in some cases)	Clean and treat environment, remove old litter, prevent wild bird access, use flea traps
Mites (e.g., Red mite- *Dermanyssus gallinae, Ornithonyssus sylviarum,* Deplumming itch- *Cnimidocoptes gallinae,* Scaly leg mite- *Cnimidocoptes mutans*)	Itching, feather loss, scabs on skin, anemia, reduced egg production	Skin scrapings, finding mites on skin or in the environment, observing mite activity at night	Permethrin sprays, ivermectin, amitraz, sulfur powder	Regular cleaning of coops, use of diatomaceous earth, sealing cracks and crevices, treating birds and environment simultaneously
Ticks (e.g., *Argas persicus*)	Anaemia, weakness, decreased egg production, irritation, tick paralysis	Visual inspection, finding ticks on skin, especially under wings and around the neck	Pyrethroid sprays, ivermectin, amitraz	Tick control in the environment, treating bedding and roosts, limiting exposure to wild birds, use of acaricides

Conclusion

Parasitic diseases in poultry can lead to significant health issues and economic losses if not managed properly. Early diagnosis, effective treatment, and stringent preventive measures are crucial in controlling these diseases. By maintaining good biosecurity, regular health monitoring, and proper management practices, poultry farmers can mitigate the impact of parasitic infections and ensure the health and productivity of their flocks.

References

Ascaridiasis in Poultry." In Diseases of Poultry, 13th Edition, edited by D.E. Swayne, John Wiley & Sons 2013, pp. 921-928.

Belete, A., Addis, M. and Ayele, M. 2016. Review on major gastrointestinal parasites that affect chickens. Journal of Biology, Agriculture and Healthcare, 6(11), pp.11-21.

Britez, J.D., Rodriguez, A.E., Di Ciaccio, L., Marugán-Hernandez, V. and Tomazic, M.L. 2023. What do we know about surface proteins of chicken parasites Eimeria. Life, 13(6), p.1295.

Capillariasis in Poultry." In Veterinary Parasitology, 2nd Edition, edited by M.W. Dryden, Elsevier, 2016, pp. 534-540.

Chantima, K. and Wongsawad, C. 2016. Echinostomarevolutum in domestic chickens: developmental larval stages and fecundity of an intestinal trematode.

DEVANEY, J.A. 1986. Ectoparasites. Poultry Science, 65(4), pp.649-656.

Dubey, J.P., Calero-Bernal, R., Rosenthal, B.M. Speer, C.A. and Fayer, R., 2015. Sarcocystosis of animals and humans. CRC Press.

Dubey, J.P.2016. Toxoplasmosis of animals and humans. CRC press.

Fayer, R. and Xiao, L. eds., 2007. Cryptosporidium and cryptosporidiosis. CRC press.

Forrester, D.J. and Foster, G.W. 2008. Trichomonosis. Parasitic diseases of wild birds, pp.120-153.

Herman, C.M. - "Tetrameres in Poultry." In Diseases of Poultry, 12th Edition, edited by Y.M. Saif, Iowa State University Press, 2003, pp. 1025-1030.

https://ndvsu.org/images/StudyMaterials/Parasitology/coccidiosis-in-poultry-ppt.pdf

https://www.farmhealthonline.com/en/disease-management/poultry-diseases/helminthiasis/

Johannes, K. 1996. Parasitic infections of domestic animals. A Diagnostic Manual, pp.24-27.

McDougald, J.M. - "Acuariasis in Poultry." In Poultry Diseases, 6th Edition, edited by F.T.W. Jordan, Bailliere Tindall, 2001, pp. 478-482.

McDougald, J.M. - "Heterakiasis in Poultry." In Poultry Diseases, 7th Edition, edited by M.E. Pattison, Elsevier, 2008, pp. 233-238.

McDougald, L.R., 2005. Blackhead disease (histomoniasis) in poultry: a critical review. Avian diseases, 49(4), pp.462-476.

Morishita, T.Y. and Schaul, J.C. 2007. Parasites of birds.

Peek, H.W. and Landman, W.J.M. 2011. Coccidiosis in poultry: anticoccidial products, vaccines and other prevention strategies. Veterinary quarterly, 31(3), pp.143-161.

Pote, K.A. - "Syngamiasis in Poultry." In Veterinary Parasitology, 2nd Edition, edited by M.W. Dryden, Elsevier, 2016, pp. 545-550.

Sehgal, R.N., Valki nas, G., Iezhova, T.A. and Smith, T.B. 2006. Blood parasites of chickens in Uganda and Cameroon with molecular descriptions of Leucocytozoonschoutedeni and Trypanosoma gallinarum. Journal of Parasitology, 92(6), pp.1336-1343.

CHAPTER-10

NUTRITIONAL AND METABOLIC DISEASES OF POULTRY

[1*]Yora Reena, [2]Sachin Tripura, [3]Vinay VV and [4]Piyali Kuri

[1] Assistant Professor in Animal Nutrition Division, Institute of Veterinary Science & Animal Husbandry, SOA (DU), Bhubaneswar, Odisha

[2] Assistant Professor cum Junior Scientist in Department of ILFC (Animal Nutrition) CVSc & AH, Birsa Agricultural University, Ranchi

[3] Assistant Professor, Department of Animal Nutrition, Veterinary College, Gadag, KVAFSU, Karnataka

[4] Assistant Professor in Animal Physiology Division, Institute of Veterinary Science & Animal Husbandry, SOA (DU), Bhubaneswar, Odisha

**Corresponding author: yorareen@gmail.com*

Abstract

Poultry production necessitates a diet rich in essential nutrients to ensure optimal health, growth, and productivity. These nutrients include water, proteins, amino acids, carbohydrates, fats, vitamins, and minerals. Imbalances, whether deficiencies or excesses, can lead to metabolic disorders, significantly impacting poultry health. Key issues include Fatty Liver and Kidney Syndrome (FLKS), Cage layer fatigue, Tibial dyschondroplasia (TD) and Ascites etc. Additionally, acid-base imbalances, fibre deficiencies, and disruptions in enzyme function highlight the importance of precise dietary management. Understanding and addressing these nutritional needs and disorders is crucial for maintaining a healthy and productive poultry flock.

Keywords: Metabolic disorders, Fatty Liver and Kidney Syndrome, Tibial dyschondroplasia

Introduction

Poultry plays a significant role both in India and globally, contributing to food security, nutrition, and economic development. In 2024, the world's production of poultry is projected to amount to approximately 138.75 million metric tons, up from about 115 million metric tons in 2016. Poultry farming plays a crucial role in socio-economic development by providing employment, enhancing food security, supporting rural economies, and driving innovation. Poultry is now the most consumed type of meat globally, surpassing beef and pork in many regions. This trend is expected to continue as populations grow and dietary preferences evolve. Proper nutrition is essential for the health of pullets, the quality of egg production in laying hens, and the growth of broilers. But due to huge demand, poultry production undergoes intensive production, which entails growing large numbers of birds in limited spaces (Hafez and Attia 2020) and failure to meet nutrient requirement of poultry birds. Deficiencies of nutrients, imbalances and interactions among nutrients are reflected in the performance of the birds leading to nutritional and metabolic diseases in poultry birds. Addressing nutritional and metabolic diseases in poultry is vital for ensuring animal welfare, improving productivity, enhancing economic viability, and supporting global food security. By prioritizing bird health through proper nutrition and management, farmers can achieve better outcomes for their flocks and contribute positively to the agricultural sector.

Here we have discussed common nutrient deficiencies and metabolic disorders faced by poultry farmers and industries, along with their control and prevention strategies.

Essential Nutrients in Poultry Production

Poultry require a balanced diet containing at least 36 essential nutrients. Common feed ingredients often lack many of these nutrients, so it's necessary to add purified sources to their diet (Saif, Y.M. 2009). Mistakes in diet formulation or milling can lead to nutrient deficiencies or toxicities. Even slight shortages can result in poor growth, weakened disease resistance, lower egg production, or reduced hatchability. Some nutrient needs are still not fully understood. Key nutrients for poultry include water, proteins and amino acids, carbohydrates, fats, vitamins, and essential minerals. Table 1 lists common nutritional disorders in poultry and their associated nutrient deficiencies.

Table 1. Nutritional diseases related disorders in Poultry

Nutritional Disorder	Deficient nutrients
I. General	
1. Malnutrition & reduced feed intake	Amino acids/protein, minerals and vitamins.
2. Stunted growth and lack of vigour	do
3. Diarrhoea	Niacin, riboflavin,
II. Feather Abnormalities	
1. Uneven feather growth, ruffled feathers, abnormally long feather	Amino acids imbalances
2. Frizzled and rough feathers	Calcium and zinc minerals, vit. niacin and pantothenic acid
III. Skin lesions	
1. Dermatitis, scabs and crust around eyes and vents	Biotin and Pantothenic acid
2. Ulcerated foot pad, haemorrhagic cracks and necrotic toes	Biotin
3. Lesions around eyes, eyelids stuck together	Vit.A
IV. Leg and Bone deformities	
1. Soft curved and weak bones	Vit. D., Ca and P
2. Bowed legs, enlarged hocks	Zinc or Niacin
3. Tibiotarsi joint enlarged	Niacin
4. Slipped tendon or perosis	Zinc, Manganese, Choline, Folacin, biotin and Vit. B-12
5. Shortening or thickening of leg bones in chicks	Manganese or zinc
6. Curled toes inwards in chicks	Riboflavin
7. Weak legs, Penguin like Squat	Vit. D3
V. Muscle disorders	
1. Gizzard myopathy	Selenium and, or Vit. E.
2. Muscular dystrophy, degeneration in skeletal muscle.	do
VI. Blood, vascular disorders	
1. Hypochromic anaemia	
(1) Macrocytic	Folacin
(ii) Microcytic	Iron or Copper
2. Aortic rupture causing internal haemorrhages	Vit. K
3. Enlarged heart.	Copper
4. Exudative diathesis	Selenium or Vit. E.

Nutritional Disorder	Deficient nutrients
VII. Nervous disorders in chicks	
1. Head drawn back/polyneuritis	Thiamin
2. Ataxia in growers	Vit. A
3. Hyperexcitability & convulsions	Vit. B6
4. Ataxia-with star-gazing posture like in thiamine deficiency	Manganese
5. Fright reaction and tetanic spasm.	Chlorides
6. 'Curled-toe' paralysis.	Riboflavin
7. Encephalomalacia, head retraction, with tetanic spasm, legs outstretched.	Vit. E.
VIII. Mouth cavity disorders	
1. Tongue deformity	Leucine, Isoleucine, Phenylalanine
2. Inflammation of tongue and mouth cavity	Niacin

Source: Extracted from *Principles & practices of animal nutrition* **book by Jagdish Prasad (2019).**

Impact of Nutrients on Metabolism

Nutrients are vital for metabolism, with numerous micronutrients actively participating in enzyme and hormone functions. Deficiencies or excesses of specific nutrients can lead to nutritional, toxic, or metabolic disorders.

Vitamin D3 Deficiency and Toxicity

A lack of vitamin D3 can cause rickets, a painful skeletal disorder, often classified as a metabolic condition in poultry. On the other hand, excessive vitamin D can also disrupt the endocrine system.

Role of Vitamin C

Vitamin C has a specific impact on metabolism by reducing the metabolic rate (Julian, 2000).

Free Radical-Induced Diseases- Conditions like encephalomalacia, nutritional muscular dystrophy, and transudative diathesis may be linked to an excess of polyunsaturated fatty acids or a deficiency of vitamin E or selenium. These are considered nutritional metabolic disorders. Free radicals can damage cell membranes, leading to cell death.

Selenium and Vitamin E Deficiency

In birds deficient in selenium and vitamin E, muscle disorders are more prominent in skeletal, gizzard, or intestinal muscles than in the heart. However, turkeys and ducks can develop heart lesions, visible as white streaks or patches, linked to hydropericardium. These lesions result from muscle fiber degeneration and mineralization. Encephalomalacia involves capillary damage, fibrin clots, hemorrhage, and ischemic necrosis.

Minerals and Vitamin B complex deficiency

Iodine is essential for producing thyroid hormones, which regulate metabolism, growth, and development in the most body tissues and organs. Phosphorus, zinc, chromium, manganese, copper and other mineral and vitamins, particularly the vitamin B complex, are required nutrients for enzyme function. The clinical signs of **riboflavin deficiency** (peripheral neuropathy and paralysis) are caused by an enzyme deficiency.

Acid-Base imbalance

The acid-base balance of poultry feed, typically characterized by a low pH, influences various aspects of poultry health, including egg-shell quality, kidney function, and the occurrence of leg problems in broilers. This imbalance is often due to high levels of chloride, a strong ion, or sulphate. However, other negative ions (anions) can also contribute to an increased anion/cation ratio.

Fiber Deficiency

A lack of fiber in poultry diets can lead to underdevelopment of the gizzard muscle, resulting in a backup of feed in the proventriculus and dilation of the preventriculus, especially when daily feed intake is high.

Metabolic disorders in poultry:

1. Fatty Liver and Kidney Syndrome (FLKS)

This metabolic disease in 2-3 weekold broiler chicks is caused by biotin deficiency, leading to impaired liver function and increased fat storage. It results from low activity of the biotin-dependent enzyme pyruvate carboxylase, causing hypoglycaemia and sudden onset of symptoms like loss of appetite, lethargy, and uncoordinated behaviour. Mortality rates range from 5% to 35%. Necropsy reveals enlarged, pale, and fatty liver and kidneys. Adequate dietary biotin can prevent the disease, which is now rare in broilers on commercial diets.

2. Fatty Liver Haemorrhagic Syndrome (FLHS)

It is a condition that affects caged birds (laying hens and breeders) on high-energy rations, and mostly prevalent in the summer. There is increased accumulation of fat in the liver. Pullets carrying an excess of body fat are more prone for this condition. Haemorrhage occurs from a ruptured liver. Fatty liver disease affects the bird's calcium metabolism, thus compromising skeletal integrity and eggshell quality.

Etiology- Low protein, high energy rationsand limited exercise are linked to the illness; *Rations* with amino acid imbalance or deficiency; Diets with imbalance amino acid and low lipotropic factors such as choline, methionine and vitamin B12. Poultry fed moulds growth feed and rapeseed meal that contains high level erucic acid.

Clinical Signs & symptoms:

- FLHS birds shows no indications of illness.
- Sudden death along with fat accumulation and liver bleeding in birds

Control and Prevention

- When FLHS identified in birds, reduced the energy content of the diets.
- Increase the level of dietary protein by 1 to 2%
- Supplementation of 50 g of copper sulfate, 500 g of choline, 3 mg of vitamin B12, 5000 IU vitamin E and 500 g of DL-methionine per 1000 kg of feed may help.

3. Gout (Urate deposition)

Visceral or articular gout, also known as urolithiasis, is a common consequence of kidney failure. When the kidneys fail, they cannot effectively clear uric acid from the blood, resulting in hyperuricemia and the formation of insoluble monosodium urates within the kidneys or on the surfaces of other visceral organs. These urates appear as semisolid, white chalky deposits and should be differentiated from fibrinoid or purulent inflammatory exudates caused by infectious diseases such as synovitis, peritonitis, perihepatitis, and pericarditis. Besides visceral organs, these urates can also be found in the synovial fluid and tendon sheaths of various joints, particularly the hock joint.

4. Cage Layer Fatigue

High-producing laying hens kept in cages sometimes exhibit paralysis around peak egg production. This condition is caused by a vertebral fracture that subsequently affects the spinal cord. The fracture results from impaired calcium flux due to the high calcium output in eggshell formation. As medullary bone reserves deplete, the bird utilizes cortical bone as a calcium source for the eggshell. This condition is rarely observed in deep litter systems, suggesting that reduced activity and exercise are predisposing factors.

Prevention of Cage Layer Fatigue (CLF)

- Weight Management: Ensure pullets reach normal weight-for-age at sexual maturity.
- Dietary Adjustments: Provide a high-calcium diet (minimum 3.5% Ca) at least 14 days before the first oviposition.

5. Tibial dyschondroplasia (TD)

It occurs when prehypertrophic cartilage in the growth plate fails to become hypertrophic cartilage, leading to an abnormal cartilage mass. This condition is common in commercial meat-type poultry, especially in the proximal tibia. Small lesions cause minimal changes, but larger ones can weaken and deform bones, sometimes leading to lameness. Diagnosis requires a post-mortem exam or radiograph. TD accounts for 5-25% of lameness in broilers, and in some flocks, it can cause up to 50% of lameness.

Etiology- Dyschondroplasia arises from several factors, such as rapid growth without adequate rest, a lack of growth factors, a high anion/cation ratio in the feed, and genetic predisposition.

Prevention Strategies

1. **Dietary Adjustments:** Alkalizing the feed by replacing 1 kg of NaCl with 2 kg of $NaHCO_3$ per tonne can reduce the incidence of dyschondroplasia.
2. **Vitamin Supplementation**: Adding 5-10 µg of 1,25-dihydroxycholecalciferol per kg of feed may prevent the condition. However, since vitamin D deficiency can lead to acidosis, the pH level in the growth plate may be more important than previously understood.

6. Ascites

Metabolic diseases like ascites in broiler chickens lead to considerable economic losses in the poultry industry. This syndrome has multiple causes, including external factors such as environmental conditions and internal factors like rapid growth rates and reactive oxygen species. Ascites, also called pulmonary hypertension syndrome/ pulmonary arterial hypertension syndrome, occurs due to an imbalance between the heart's output and the pulmonary blood vessels' capacity to handle the increasing blood flow.

Common symptoms of ascites in broilers include: Swollen abdomen, difficulty breathing, bluish skin (especially around the comb and wattles), lethargy, weakness and unsteady movement and red abdominal skin.

Dietary adjustments to manage ascites in broilers-

- Reduce Sodium Levels: Lowering the sodium content in the feed can help reduce water retention and fluid accumulation.
- Increase Antioxidants: Adding antioxidants like vitamin E and selenium can help mitigate oxidative stress.
- Balanced Electrolytes: Ensuring a proper balance of electrolytes (sodium, potassium, and chloride) can help maintain fluid balance.
- Feed Restriction: Implementing controlled feeding programs to slow down growth rates can reduce the risk of ascites.

- Supplementation: Adding specific supplements like omega-3 fatty acids can improve cardiovascular health.

7. Sudden death syndrome (SDS)

Sudden death syndrome (SDS) refers to the unexpected death of healthy, fast-growing commercial broilers. These young chickens can die suddenly while standing, walking, sparring, or feeding, often experiencing a brief terminal convulsion and being found on their backs. SDS can begin as early as day 3 and persist for 8–12 weeks, with peak mortality occurring between days 9–22. This sudden death is believed to result from ventricular fibrillation and is closely linked to feed intake. SDS can be entirely prevented by limiting feed intake to 75% of what would be consumed if fed freely. Reducing energy intake by altering feed texture or density (such as using mash) or implementing management strategies like feed restriction or extended dark periods can decrease SDS-related mortality.

Conclusion

Ensuring a balanced diet with all essential nutrients is crucial for the health and efficiency of poultry production. Nutritional imbalances, deficiency or excess can cause severe metabolic disorders, impacting growth, egg production, and overall health. By carefully formulating and monitoring diets, poultry producers can prevent metabolic health issues and enhance productivity, sustaining a healthy and productive flock.

Reference

Bender, D.A., 2002. An Introduction to Nutrition and Metabolism. Taylor and Francis Ltd., London. 450 pp.

Hafez, H. M., & Attia, Y. A. (2020). Challenges to the poultry industry: current perspectives and strategic future after the COVID-19 outbreak. *Frontiers in veterinary science*, *7*, 516.

Julian, R. J. (2000). Physiological, management and environmental triggers of the ascites syndrome: a review. *Avian pathology*, *29*(6), 519-527.

Prasad, J. (2019). Poultry nutrition. In Principles & Practices of Animal Nutrition (5th Revised ed., pp. 336-353). Kalyani Publishers.

Saif, Y. M. (2009). *Diseases of poultry*. John Wiley & Sons.

CHAPTER 11

MANAGEMENT OF POULTRY DISEASES

[1*]B. Divya Sri, [2]Deepthi Chandaka and [3]Saikumar Gangitla

[1] Assistant Professor, Department of Veterinary Physiology & Biochemistry School of Veterinary & Animal Sciences, CUTM, Paralakhemundi, Odisha.
[2] Assistant Professor, Department of Livestock Farm Complex School of Veterinary & Animal Sciences, CUTM, Paralakhemundi, Odisha.
[3] Assistant Professor, Department of Veterinary Microbiology School of Veterinary & Animal Sciences, CUTM, Paralakhemundi, Odisha.
*Corresponding author-divyasri@gmail.com

Abstract

Poultry health management is a critical aspect of ensuring sustainable production. This review summarizes common poultry diseases and their prevention, focusing on infectious and non-infectious causes. Diseases result from the interaction of host, agent, and environmental factors, with management playing a significant role in disease prevention. Infectious agents include bacteria, viruses, parasites, and fungi, while non-infectious chemical and physical stressors contribute to health issues. Early signs of disease often manifest as poor appetite, huddling, and respiratory distress, which can indicate more serious conditions like Avian Influenza, Marek's Disease, or Coccidiosis. Prevention and control strategies emphasize biosecurity, vaccination, and proper management practices. Key measures include isolating sick birds, maintaining good sanitation, providing adequate ventilation, and ensuring feed and water quality. The document also highlights that not all health problems are caused by infectious agents, urging poultry managers to first address non-infectious conditions such as feed and space issues. Maintaining strict biosecurity, monitoring flock health, and seeking expert advice are essential for minimizing the risk of disease outbreaks in poultry.

Keywords: Poultry health, Disease prevention, Infectious agents, Non-infectious factors, Vaccination, Poultry management, Disease control, Host factors, Environmental factors.

Introduction

Effective poultry health management ensures optimal production and minimizes disease-related economic losses. Disease is any condition that disrupts or alters normal body functions, leading to deviations from health. In poultry farming, early detection of disease signs such as poor appetite, respiratory distress, and abnormal behavior is crucial for prompt intervention. Preventive measures such as strict biosecurity protocols, vaccination programs, and environmental factors like feed and ventilation management are necessary to protect poultry from these threats. Understanding the interaction between host, agent, and environmental factors is key to preventing disease outbreaks.

Poultry diseases arise due to the interaction of three primary factors:

1. **Agent Factors**

- **Infectious Agents**:
    - Bacteria
    - Viruses
    - Parasites
    - Fungi
- **Non-infectious Agents**:
    - Chemical
    - Physical
    - Imbalance or deficiency of vitamins and minerals
    - Toxins

2. **Host Factors**

- Breed
- Age
- Sex
- Immune status

3. **Management and Environmental Factors**

- Feed quality
- Lighting program
- Air quality and ventilation
- Water quality

- Space (floor or cage, feeder, drinker)
- Sanitation
- Weather conditions
- Season
- Geographic location

Impact of Breed, Age, Sex, and Immune Status on Poultry Diseases

1. **Breed**

- **Impact**: Varies in resistance and susceptibility to diseases due to genetic factors.
- **Examples**:
  - » Broiler breeds are more prone to respiratory diseases.
  - » Heritage breeds tend to have better resistance compared to commercial hybrids.

2. **Age**

- **Impact**: Affects immune response; younger birds are more vulnerable to infections.
- **Examples**:
  - » Chicks under three weeks are highly susceptible to coccidiosis and salmonellosis.
  - » Adult birds may experience age-related diseases, such as reproductive issues in layers.

3. **Sex**

- **Impact**: Influences disease susceptibility and symptom expression due to hormonal differences.
- **Examples**:
  - » Male birds (roosters) may suffer from stress-related diseases due to aggression.
  - » Females may face egg production-related issues, like egg binding.
  - » **Immune Status**
- **Impact**: Critical for resisting infections; a strong immune system enhances protection.
- **Examples**:
  - » Vaccinated birds generally show better immunity.
  - » Stressors like overcrowding and poor nutrition can weaken the immune system, increasing vulnerability to diseases such as Newcastle disease and avian influenza

4. **Feed Quality:** High-quality feed is essential for optimal growth, immune function, and overall health. Poor feed quality can lead to nutritional deficiencies, making birds more susceptible to infections.
5. **Lighting Program:** Proper lighting influences growth, behavior, and reproductive performance. Inadequate or excessive light can stress birds and affect their immune response.
6. **Air Quality and Ventilation:** Poor air quality (e.g., high ammonia levels, dust, and pathogens) can lead to respiratory diseases and overall stress in poultry. Adequate ventilation is crucial for maintaining a healthy environment.
7. **Water Quality:** Clean, fresh water is vital for hydration and overall health. Contaminated water can lead to diseases and stress, impacting growth and egg production.
8. **Space (Floor or Cage, Feeder, Drinker):** Crowding can lead to stress, aggression, and increased transmission of diseases. Adequate movement, feeding, and drinking space is essential for reducing stress and maintaining health.
9. **Sanitation:** Proper sanitation practices are critical for preventing the spread of diseases. Contaminated surfaces and equipment can harbor pathogens that affect poultry health.
10. **Weather Conditions:** Extreme weather (hot or cold) can lead to stress, reduced feed intake, and increased susceptibility to disease. Temperature fluctuations can also affect respiratory health.
11. **Season:** Seasonal changes can influence disease prevalence, with some diseases more common in specific seasons (e.g., respiratory diseases in winter).
12. **Geographic Location:** The geographic location affects disease exposure, local pathogens, and environmental conditions (e.g., humidity, temperature). Different areas may have specific disease challenges.

Disease Control Management Practices

1. **Biosecurity Measures**

- Implement strict biosecurity protocols to prevent the introduction and spread of pathogens. This includes controlling access to poultry facilities, disinfecting equipment, and managing traffic flow to minimize contamination.

2. **Vaccination Programs**

- Establish a comprehensive vaccination schedule tailored to the specific diseases prevalent in the region and the flock's needs. Regularly review and update vaccination protocols based on the latest guidelines and disease outbreaks.

Table 1. Vaccination schedule for various poultry disease

Age	Vaccine	Disease Prevented
Day 1	Marek's Disease Vaccine	Marek's Disease
Day 7-10	Newcastle Disease Vaccine (Live)	Newcastle Disease
Day 12-14	Infectious Bursal Disease Vaccine	Gumboro Disease (Infectious Bursal Disease)
Day 18-21	Coccidiosis Vaccine	Coccidiosis
Week 5-6	Newcastle Disease (Booster)	Newcastle Disease
Week 8-10	Infectious Bronchitis Vaccine	Infectious Bronchitis
Week 12-16	Fowl Pox Vaccine	Fowl Pox
Week 18-20	Avian Influenza Vaccine	Avian Influenza
At Lay	Newcastle Disease (Booster)	Newcastle Disease
At Lay	Infectious Bronchitis (if indicated)	Infectious Bronchitis
Annually	Fowl Pox (if at risk)	Fowl Pox

3. **Regular Health Monitoring**

- Conduct routine health checks and monitoring of flock behavior, weight gain, egg production, and overall health. Early detection of any health issues allows for prompt intervention.

4. **Nutritional Management**

- Provide balanced and high-quality nutrition to support immune function and overall health. Adjust feed formulations based on the specific needs of different age groups or breeds.

5. **Environmental Management**

- Ensure optimal housing conditions by maintaining appropriate temperature, humidity, and ventilation. Regularly clean and disinfect housing to reduce pathogen loads.

6. **Sanitation Practices**

- Implement rigorous cleaning and disinfection protocols for all equipment, housing, and feeding areas. Use appropriate disinfectants and follow manufacturer guidelines for effectiveness.

7. **Stress Reduction**

- Minimize stressors by providing adequate space, reducing overcrowding, and maintaining a stable environment. Address factors such as lighting, handling, and transportation to promote flock well-being.

8. **Water Quality Management**

- Ensure access to clean and fresh water at all times. Regularly test water sources for contaminants and maintain proper sanitation of water delivery systems.

9. **Quarantine and Isolation**
   - Isolate new birds or those showing signs of illness from the main flock to prevent disease spread. Implement a quarantine period for all incoming birds to monitor for any signs of infection.
10. **Record Keeping**
   - Maintain detailed records of flock health, vaccination schedules, feed formulations, and any disease outbreaks. Analyzing this data can help identify trends and improve future management practices.

Conclusion:

In conclusion, effective poultry health management is crucial for sustainable production and minimizing disease-related losses. Key strategies include strict biosecurity, sanitation, vaccination programs, and regular flock health monitoring. Providing optimal nutrition and environmental conditions helps reduce stress and disease risk. Understanding factors like breed, age, and immune status allows for targeted management practices that enhance flock health. A proactive, integrated approach to health management ensures healthier birds and more profitable production.

References

Belete, A., Addis, M. and Ayele, M. 2016. Review on major gastrointestinal parasites that affect chickens. Journal of Biology, Agriculture and Healthcare, 6(11), pp.11-21

OIE (2024). Manual of Diagnostic Tests and Vaccines for Terrestrial Animals,(13th ed.).

Prasad, J. (2019). Poultry nutrition. In Principles & Practices of Animal Nutrition (5th Revised ed., pp. 336-353). Kalyani Publishers.

Saif, Y. M. (2009). *Diseases of poultry*. John Wiley & Sons.

Shane, S. M., & Association, A. S. (2005). *Handbook on Poultry Diseases*.

CHAPTER 12

POULTRY MEAT PROCESSING AND VALUE ADDED BY PRODUCTS OF POULTRY INDUSTRY

*[1]M. Bindu Pushpa, [2]Anisetty Shilpa Sree and [3]Surita Majumder

[1] Assistant Professor, Department of Veterinary Public Health, Institute of Veterinary Science & Animal Husbandry, SOA DU, Bhubaneswar, Odisha

[2] Assistant Professor, Department of Livestock products and Technology, Institute of Veterinary Science & Animal Husbandry, SOA DU, Bhubaneswar, Odisha

[3] Assistant Professor, Department of Animal Genetics and Breeding, Institute of Veterinary Science & Animal Husbandry, SOA DU, Bhubaneswar, Odisha

*Corresponding author-bindupushpa1221@gmail.com

Abstract

Poultry meat processing is a vital sector of the food industry, offering affordable and high-quality protein sources worldwide. The process involves slaughtering, deboning and further processing of poultry into cuts or products like sausages, nuggets and deli meats. Alongside primary meat production, the industry generates substantial by-products, such as feathers, blood, bones, and offal, which can be transformed into value-added products.By-products from poultry processing can be repurposed into a range of products, including animal feed, biofuels, organic fertilizers and collagen-based products. Feathers, rich in keratin, can be processed into biodegradable plastics or used as soil conditioners. Poultry fat is used in biodiesel production, and eggshells serve as calcium supplements. The integration of innovative technologies, such as bio-processing and microbial fermentation, enhances the sustainability of poultry production, transforming waste into valuable resources and contributing to a circular economy while reducing environmental impacts.

Keywords: Poultry processing, Emulsification, Restructured, Byproducts, feathers, blood

Introduction

Poultry meat processing involves treatments, such as salting, that significantly alter the meat's chemical and physical properties from its original form. The key advantages of processing include value addition, masking undesirable attributes like odor and color, minimizing waste for both consumers and processors, ensuring uniformity in texture, color, and fat distribution, and providing ready-to-eat, healthy, and convenient food products. Poultry, like other livestock, provides not only meat but a wide range of value-added byproducts, classified as edible, inedible and pharmaceutical.

Edible byproducts from poultry include organs like the liver and heart, which are used in food products, while inedible byproducts such as feathers and fat can be repurposed for industrial applications like biodiesel, glues and biodegradable plastics. Poultry byproducts also have pharmaceutical applications, with enzymes and other components used in various medicinal products. Modern technologies and efficient waste utilization methods transform poultry byproducts into valuable resources, enhancing sustainability and minimizing environmental impact.

Poultry meat processing

Processed meat products from spent animals and birds, which are typically tough, can be efficiently utilized in various categories such as emulsion-based, restructured, enrobed, fermented, cured and smoked products. These products offer convenience, enhanced taste, and extended shelf life, often requiring minimal capital investment for processing technologies. Processing of meat involves various treatments that alter the chemical and physical properties of meat, such as emulsification, preblending, hot processing and cooking. Emulsification is the mixture of immiscible liquids, where fat droplets are dispersed in a continuous water phase. In meat emulsions, proteins act as emulsifiers, and maintaining low temperatures is crucial to avoid fat melting and protein denaturation.

Meat cooking methods include dry heat (broiling, roasting, frying), moist heat, and microwave cooking, each suited for different cuts of meat. Processing techniques are classified into categories like cured and smoked meats, sausages, restructured meat, and shelf-stable products. These methods enhance flavour, texture and shelf life, adding value to meat products. Curing involves the distribution of salt and curing agents throughout large cuts of meat, followed by smoking to preserve the product and enhance its aroma and flavour. Commonly cured and smoked products include hams and beef cuts

Emulsion-Based Meat Products: Emulsion-type products, like meatballs, patties, and nuggets, are popular due to their ease of preparation. Lean muscle tissue, fat, water and salt are mixed to form a stable emulsion, preventing fat and water separation during cooking. These products are versatile, ready-to-eat and commonly used in sandwiches and snacks.

Restructured Meat Products: Made from less expensive cuts or meat trimmings, restructured meat products involve binding small pieces of meat to form larger cuts. They offer sensory qualities similar to intact steaks and are marketed in precooked or frozen forms. Examples include restructured chicken slices, blocks, and meat rolls.

Fermented Meat Products: Fermentation is a low-energy preservation method that enhances meat flavour, colour, tenderness and microbiological safety. Microorganisms are used to lower the pH of the meat, producing items like dry and semi-dry sausages, which rely on microbial fermentation and dehydration.

Meat extenders like soy, starch, and gums stabilize emulsions and reduce costs, making meat products more affordable. Preblending enhances protein extraction and allows better control over meat composition. Hot processing, where the carcass is processed shortly after slaughter, improves product quality and cooking yield while reducing energy and labor costs.

Fig 1. Value added by products of poultry industry (Source:http://shutterstock.com)

The poultry meat industry generates significant by-products like blood, bones, meat scraps, skin, fatty tissues, and viscera, posing high ecological costs for treatment and disposal. These by-products, commonly used for livestock feed and organic fertilizers, are rich in proteins, fats, carbohydrates, vitamins, and minerals, making them suitable for deriving functional ingredients such as bioactive peptides. Proteins in poultry by-products are hydrolysed into peptides, which have potential health benefits, including antihypertensive, antioxidant, antidiabetic, antimicrobialand anti-inflammatory properties.

Enzymatic hydrolysis is often used to obtain these peptides with high yields and improved functional properties like solubility and emulsification. However, bitterness from protein hydrolysates can limit their application, leading to methods like exopeptidase treatment and encapsulation to improve taste. Peptides derived from poultry by-products, typically consisting of 2-30 amino acids and having a molecular weight of 400-2000 Da, show great potential as bioactive compounds with health benefits. Additionally, the poultry industry produces a wide range of waste products that can be effectively repurposed into valuable by-product.

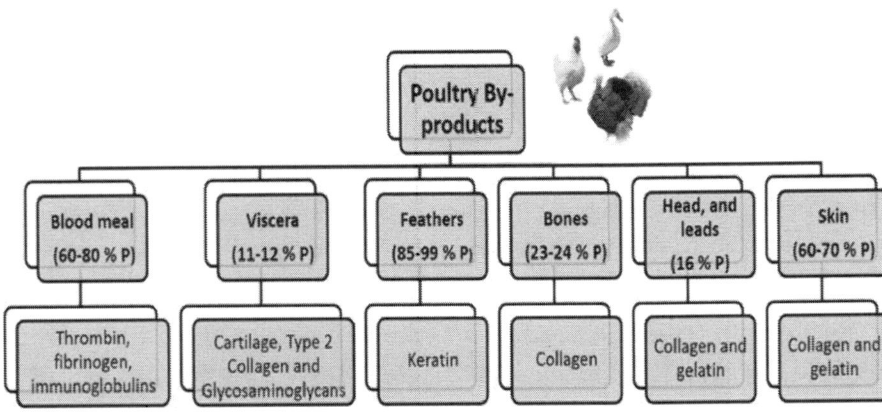

Edible poultry by products

Edible poultry by-products, such as dressed whole chicken carcasses and giblets like liver, heart, gizzard, and neck, have strong market value and are often sold packaged inside the cavity of the carcass. Minced skin, gizzards, and hearts are commonly used in chicken product emulsions. Ova collected during the dressing of layer chickens are incorporated into sausage formulations, patties and kebabs. Additionally, chicken soup, a nutritious and low-fat food, is a popular option.

Inedible poultry by products

Blood: Accounting for 3-3.7% of a bird's body weight, blood is rich in haemoproteins and iron, and can be used in edible products like blood meals, sausages, and cakes, as well as non-edible applications such as fertilizers, animal feed and emulsifiers.

Poultry Litter: Composed of urine and faeces, poultry litter poses environmental risks but can be used as organic fertilizer, livestock feed, or processed through anaerobic digestion to produce biogas and nutrient-rich manure. Composting also produces pathogen-free fertilizer.

Feathers: Comprising 90% keratin, feathers are a significant waste product. They are processed into feather meal for biofertilizers and other uses such as insulation material, compost, and decorative items.

Hatchery Waste: Including eggshells, infertile eggs, and dead chicks, hatchery waste can be processed into calcium-rich by-products for food and agricultural uses.

Offal: Poultry offal, including organs, bones, and blood, can be repurposed for animal feed or human consumption and is used in non-edible applications like fertilizers and stabilizers.

Eggshells & Membranes: Eggshells, composed primarily of calcium carbonate, can be used in biodiesel production, fertilizers and calcium supplements. Eggshell membranes, rich in collagen, have medical and industrial applications.

Spent Hens: Spent hens provide meat and protein, used in food products and animal feed, with by-products also utilized in adhesives and mutton nuggets.

Conclusion

Waste from the poultry industry, if not properly managed, can become a significant environmental pollutant. However, poultry waste, including feathers, manure, eggshells and bedding material, can be repurposed into valuable products such as feed, fertilizer, biofuel and biodegradable plastic. Feathers, for example, can be processed naturally or with organisms to enhance their nutritional value and used as compost or soil conditioners. Poultry litter can also generate methane gas, which can be converted into energy.

The industry produces large amounts of waste, and improper disposal can lead to air, water, and land pollution, as well as the spread of pathogenic microbes. Traditional waste disposal methods, combined with modern technologies like microbial and enzymatic processes, can help manage waste more sustainably. Innovations like the use of keratinolytic bacteria for feather degradation, bioprocessing of hatchery waste into feed or fuel and the conversion of eggshells into useful biomaterials are crucial steps toward achieving zero-waste goals.

Raising awareness among poultry farmers, especially small-scale operations and promoting the use of advanced waste management technologies is essential. Policymakers also play a key role in supporting these efforts through education, policies and subsidies for sustainable practices.

References

Anjaneyalu, A.S.R., Muthukumar, M. and PrinceDevadason, I., 2012. Processing technologies of value-added chicken products for employment generation. *IPSACON, pg*, pp.116-121.

Brar, P.S., Mehta, N., Singh, A., Sivakumar, S. and Phand, S., 2021. Value addition of Milk and meat: A push to entrepreneurship [E-book]. *Hyderabad: Guru Angad Dev Veterinary and Animal Sciences University, Ludhiana & National Institute of Agricultural Extension Management, Hyderabad, India.*

Karuppannan, S.K., Dowlath, M.J.H., Raiyaan, G.D., Rajadesingu, S. and Arunachalam, K.D., 2021. Application of poultry industry waste in producing value-added products—A review. *Concepts of advanced zero waste tools*. 91-121.

Ojha, S., Gaikwad, S.T., Suthar, T. and Gavane, A., 2020. Microbial bioconversion of poultry waste: value added products. *Indian J. Pure Appl. Biosci.*, 8: 165-173.

Romero-Garay, M.G., Montalvo-González, E., Hernández-González, C., Soto-Domínguez, A., Becerra-Verdín, E.M. and García-Magaña, M.D.L., 2022. Bioactivity of peptides obtained from poultry by-products: A review. *Food Chemistry: X*, 13:100181.

Sharma, B.D., 1999. *Meat and meat products technology: (including poultry products technology)*. Jaypee Brothers Medical Publishers Limited

CHAPTER 13

POULTRY-FACTS, MYTHS AND MISCONCEPTIONS

[1*]V.N. Sahana, [2]Robin Tiwari and [3]Ilayaraja

[1] PhD Scholar, Division of Animal Genetics and Breeding, ICAR-NDRI, Karnal, Haryana

[2] MVSc Scholar, Department of Veterinary Medicine, OUAT, Bhubaneswar

[3] MVSc Scholar, Division of Animal Genetics and Breeding, ICAR-NDRI, Karnal, Haryana

Corresponding author-sahanavn01@gmail.com

Abstract

The poultry industry has emerged as one of the fastest-growing and most adaptable sectors of livestock farming, driven by increasing consumer demand and significant advancements in genetics, nutrition, housing and management practices. Despite notable improvements in health and welfare traits of poultry over the past few decades, the industry faces challenges from widespread myths and misconceptions regarding poultry eggs and meat. This chapter addresses prevalent myths, such as misconceptions about egg production, dietary cholesterol, and the safety of broiler chicken consumption. It highlights the economic impacts of misinformation on the poultry sector, including decreased consumer demand and financial strain on farmers. Furthermore, effective strategies for combating these myths through education and community engagement are discussed. Ultimately, the chapter emphasizes the importance of accurate information to promote the nutritional benefits of poultry products and to support the industry's growth while ensuring consumer safety and awareness.

Keywords: Poultry sector, Egg production, Myths and Misconceptions, Broiler chicken

Introduction

The Poultry sector is possibly the fastest growing and most flexible of all livestock sectors. It has undergone rapid growth and diversification over few decades, driven by strong consumer demand. This expansion has occurred across countries of varying income levels, leading to consolidation and globalization. Advancements in genetics, nutrition, housing and management have greatly benefited the poultry industry. Today, geneticists focus on welfare and health traits in their breeding programs, leading to modern poultry being healthier than they were 30 years ago.

Worldwide, production of poultry meat and eggs has increased consistently over the years. For example, the age for commercial broiler chicken to reach a market weight of 2 kg has steadily decreased from 63 days in 1976 to 35 days in 2009 and in 2018, only 41 days were required to reach alive market weight of 2.26 kg and the efficiency of converting feed into poultry products also continues to improve.

In developing nations poultry has a direct positive impact on the well-being of the socially marginalized people by being a source of livelihood, employment and income especially for womenfolk under low input and offering high returns in a short period of time. The poultry sector in India has witnessed a phenomenal transformation in recent decades, propelling the nation to become the world's third-largest producer of eggs and the eighth in terms of total meat production. The total egg production in the country is 138.38 billion numbers during 2022-23 (BAHS, 2023). Broiler chickens, which grow from a hatch weight of 65 grams to approximately 1.5-2 kg within six weeks, are a crucial part of this industry due to their high protein content and feed conversion efficiency.

However, this growth of the poultry industry was not smooth and easy and had several impediments on its way to its today's position. Even today one can observe a lot of misconceptions and myths surrounding poultry eggs and chicken consumption. Many nutrition and health advisory bodies may have been influenced by several recent scientific papers, which have dispelled myths around eggs and cholesterol.

Beliefs are one of the several factors that influence our eating habits, forming a foundation from set of concepts accepted as truth by an individual. Food myths are beliefs about nutrition, often contradicting established scientific evidence but nevertheless believed to be true. Many of these myths stem from misinformation, and the lack of reliable nutritional knowledge. Misperceptions about specific foods can arise from misunderstandings or misinterpretations of available information.

Henceforth listing the actual facts about it is paramount to enhance our knowledge and this chapter explores the major myths and misconceptions prevalent about poultry eggs and meat.

Myths on egg	Myths on broiler meat
All eggs are produced only upon mating between male and female birds	Consumption of broiler meat is unhealthy as they are fed antibiotics to grow faster
Egg yolk consumption increases blood cholesterol	Broilers are injected with hormones to grow faster
Eggs are non-vegetarian and cannot be consumed by vegetarian?	Consumption of poultry meat can lead to spread of COVID-19 infection in humans
Eggs must not be consumed in summers because it will increase the heat in the body	Broilers are genetically modified organisms
Raw eggs are more proteinaceous than boiled eggs	During bird flu in country people should not take meat
All brown shelled eggs are organic	Broiler meat consumption leads to early puberty and obesity
Plastic eggs are present in market	Pregnant women and young children should not eat broilers

Myths on eggs

Eggs are produced only upon mating between male and female birds

To lay eggs, male birds (cocks) are not needed for the female birds (hens). In a way similar to humans, wherein the regular periods or cycles that causes release of ova or eggs in humans, the chicken also releases ova that gets reflected in eggs. Hens will start its egg producing cycle (18 to 72 weeks) automatically after reaching its sexual maturity without mating. These eggs laid without mating are called as table eggs. In commercial farms, only female birds are reared and the eggs laid by them are used for consumption only. Farmers only need to keep both male and female birds together when they want to produce hatching eggs and not for producing table eggs. *Therefore, for female bird to lay egg male bird is absolutely not necessary and no need of any mating at all. Whether male is present or not, female chicken lays eggs continuously.*

Eggs are non-vegetarian and cannot be consumed by vegetarian?

Eggs can be classified into two types: fertile eggs, which contain embryos, and sterile or table eggs, which do not contain embryos. Hens that are not housed with males produce unfertilized eggs that lack embryos and cannot hatch into chicks. All the eggs found in supermarkets and restaurants are sterile eggs, meaning they do not contain embryos or chicks. These eggs are table eggs and are equally considered as milk. Milk also contains animal cells same as in eggs. Fertile eggs are not available in market for consumption and only used for hatching chicks.

Egg yolk can lead to increase in cholesterol level in body

Dietary cholesterol content from egg and chicken does not affect the amount of cholesterol in the body. A medium-sized boiled egg (50 g) contains 78 kcal energy, 6.29 g protein, 0.56 g carbohydrate, and 5.3 g total fat, of which 1.6 g is saturated,

2.0 g is monounsaturated, 0.7g is polyunsaturated, and 186 mg is cholesterol (less than dietary intake of cholesterol needed by body). The 2015–2020 Dietary Guidelines for Americans recommended egg consumption as part of a healthy diet, and abolished the limit of dietary cholesterol intake of 300 mg/day that has been in the American dietary guidelines since 1968. For a healthy adult, it's safe to eat 3 eggs a day depending on how much other cholesterol is in diet.

Poultry must not be consumed in summers because it will increase the heat in the body or elder people and pregnant ladies are not supposed to eat eggs

Heat increment in the body is related to the calorific value of the food item. Eggs have lower energy content (78 Cal/egg) and egg contains a variety of minerals (calcium, iron, magnesium, phosphorus, potassium, sodium, and Zinc) and most vitamins (thiamine, riboflavin, niacin, vitamin B_6, folate, vitamin B_{12}, vitamin A, vitamin E, vitamin D, and vitamin K) except for vitamin C, which are incredibly health beneficial and can be easily digested by elderly people. Similarly, these nutrients are essential for the growing foetus for a better growth and functioning of new born baby.

Plastic eggs are present in market

The concept of plastic eggs is a significant misconception. In summer, the shells of eggs can become weak and thin due to a loss of carbonates in the birds. This occurs because the feed intake of layer birds decreases from around 110 grams to about 70-80 grams, leading to a reduced calcium intake, which is essential for shell formation. As a result, the eggs may have an elastic shell that is more susceptible to damage. Additionally, creating artificial or plastic eggs that contain all the necessary nutrients is not practically feasible; even if it were, the cost would be higher for producing, making it unlikely that anyone would pursue it.

Raw eggs are more proteinaceous than boiled eggs

Consuming raw eggs can lead to diseases transmitted by eggs, such as salmonellosis. Additionally, raw eggs contain avidin, a protein that binds to biotin, reducing its absorption and potentially leading to biotin deficiency.

All brown shelled eggs are organic

The colour of eggshells is the result of pigments being deposited during egg formation. The type of pigment depends upon the breed and is genetically determined. The eggs are organic or not depends on type of farm they are produced. Even white eggs can be organic in nature if they are reared in organic farms.

Eggs become greenish colour after boiling and are very harmful to eat.

Overcooking eggs can lead to the release of iron and sulphide, which may give the yolk a slight greenish tint. These eggs, known as hard-cooked eggs, are perfectly safe to eat.

Egg blood spot is because of embryo present in egg.

Blood spots in eggs occur due to the rupture of tiny blood vessels in the hen's ovaries or oviduct during egg formation. These spots are found in less than 1% of commercially sterile eggs and are not related to the presence of an embryo, making blood-spotted eggs completely safe to consume.

Double egg is not safe and should not be taken.

Double-yolk eggs, which contain two yolks instead of one, are generally safe to eat. They tend to weigh more and are sold at a premium price, but they are not commonly found in stores. Out of every 100 eggs laid by a hen, 4 to 5 may be double-yolk eggs, which is considered normal.

Myths on Broiler meat

Chickens are grown fast as they are fed with chemicals and antibiotics hence harmful & contain antimicrobial residue and cause antibiotic resistance in human pathogens

Chickens are raised without the continuous use of antibiotics; only specific, approved antibiotics are administered therapeutically during disease outbreaks. These antibiotics help chickens grow faster by preventing intestinal and bacterial diseases. Moreover, antibiotic use is discontinued 7-10 days before slaughter to ensure they are naturally eliminated from the birds' systems. As a result, the chances of antibiotic residues remaining in the chickens is extremely low.

Broiler chicken consumption is unhealthy as they are injected with hormones for faster growth

The use of hormones in poultry is not economically feasible and impractical as it would increase the cost of production and if done so handling of birds for hormonal injection is potentially stressful moment for the bird and would decrease flock performance. The reason behind rapid growth of birds is, the R&D work done over the course of last 2 to 3 decades in genetics &nutrition to develop chicken with increased growth and production. Scientists have been selectively cross-breeding chickens for the development of high producing broiler chicken. *Hence, it's not the impact of hormones but years of scientific research that has produced larger, faster-growing birds.* Also, they are produced from best parents selected by excellent breeding plan.

Consumption of poultry meat can lead to spread of COVID-19 infection in humans

In India, there is not a single case of chicken being infected with Coronavirus, also no scientific or medical evidence for the spread of coronavirus through consumption of poultry meat. On the other side, the practice in our Indian food culture to consume well boiled chicken gives us hygienic food and in this high temperature most of pathogens can't be survived. Also, addition of spices like ginger, garlic, spices, turmeric etc. which are having better antimicrobial and anti-oxidative property in the food helps to us to consume germ free food (chicken and eggs).

Broilers are genetically modified organisms

The progress in poultry growth, survivability, and health are the result of a combination of genetics, management practices, nutrition and the environmental conditions in which the birds are raised. It's important to highlight that, at present, broiler genetics do not involve any genetic modification or engineering. While molecular genetics is used to enhance our understanding of the birds' DNA, the genetic material itself remains unaltered. Chickens are not genetically engineered or modified; no artificial changes are made to their genetic make-up of an organism to enhance specific traits. Instead, they are selectively bred for desirable characteristics, similar to the methods used with other animals for thousands of years. This approach is a natural way to improve a breed's traits.

Broiler meat consumption leads to early puberty and obesity

This misbelief is created based on the assumption that broilers are reared by injecting hormones. Broiler chicken is a lean source of protein, containing minimal fat and no carbohydrates, making it an unlikely cause of obesity. Scientific studies, suggest that non-food factors like high-fat diets, junk food and genetic predispositions play a more significant role in these health concerns than broiler consumption.

During bird flu in country people should not take meat or egg

There is no scientific evidence that the bird flu virus can infect healthy humans through poultry products like meat and eggs. The World Health Organization confirms that it is safe to consume eggs and meat if they are properly washed and cooked. Indian cooking methods, which often involve heat and the use of antibacterial ingredients like turmeric, garlic, and ginger, effectively eliminate pathogens. As a result, these foods do not carry any viral strains responsible for bird flu, making them safe to eat.

Pregnant women and young children should not eat broilers

Broiler chicken is a high-quality source of protein and essential nutrients, including B vitamins, iron, and zinc, which are important for growth and development. When properly cooked, broiler chicken is safe to eat and can be a nutritious addition to the diets of pregnant women and young children.

Other myths

Poultry litter is a waste product, and poultry farms are a major source of pollution

The manure that comes out of chicken houses is locally produced, organic fertilizer. People forget that organic farmers can use it to grow their crops. It is certainly not a waste of a valuable resource.

Impact of myth and misconceptions on poultry industry

- Myths lead to a significant drop in demand for chicken and eggs, resulting in substantial sales and income losses for farmers and retailers and in turn also pressureon government to contain the outbreak and reassure the public.
- Misinformation can trigger panic buying of alternative protein sources, such as fish or legumes, which creates an oversupply of poultry and subsequently drives down prices.
- Poultry farmers experience severe financial distress due to falling demand, compounded by the high costs associated with maintaining their livestock during outbreaks. This situation can result in farm closures, job losses and instability in rural economies.
- As rumours discourage consumers from buying chicken and eggs, large quantities of poultry products often go unsold, leading to significant waste. This not only exacerbates financial losses for the poultry industry but also impacts overall economic health.
- The poultry industry relies on a stable supply chain for feeds, vaccines and transportation of products. The disruption caused by fear can hamper supply chains, leading to further financial strain.

Approaches to combat myths and misconceptions

To effectively combat rumours and misinformation in the poultry industry, a multi-faceted approach is essential. Targeted education campaigns by health authorities and poultry associations should focus on dispelling myths tailored to local concerns and delivered in regional languages. Collaborating with local influencers and food bloggers can help promote the safety of poultry consumption. Regular public statements from government and health agencies, disseminated via social media and traditional media, will clarify facts about bird flu and directly address misconceptions. Engaging local veterinary experts, health workers and poultry farmers in awareness sessions in rural areas can reassure communities and provide evidence-based information. Additionally, partnerships with social media platforms to track and flag misinformation, along with involvement from fact-checking agencies, can further reduce the spread of harmful myths.

Conclusion

Modern broiler chickens have been developed through selective breeding to enhance traits such as faster growth rates, higher meat yield, and improved feed efficiency. This genetic progress, along with better rearing practices like balanced diets and disease control, has significantly boosted global poultry production. Poultry meat and eggs are essential sources of high-quality protein, vitamins and minerals for millions. Therefore, it is always advisable to do thorough research and get the facts checked from the experts before believing the rumours and relying on the grapevine discussions. Teaming with many health benefits and being one of the primary sources of protein, poultry meat is advised for regular consumption. The intent here is to nourish our bodies with nutritionally healthy food without baseless fears and misconceptions

References

Agalya SP, Anandasekaran G, Satheeshkumar P. Myths and realities of broiler chicken consumption. International Journal of Veterinary Sciences and Animal Husbandry. 2024;9(3):415-416.

Alpha Ngunyale Myths: The Truth About Broiler Chickens in Tanzania. Tanzania Commercial Poultry Association

Basic Animal Husbandry Statistics (2023). Department of Animal Husbandry, Dairying and Fisheries. Ministry of Fisheries, Animal Husbandry & Dairying, Government of India. https://dahd.nic.in/sites/default/filess/BAHS2023.pdf

Food and Agriculture Organization of the United Nations. 2013. Poultry development review.

ICAR- Central Avian Research Institute. Addressing myths and rumors regarding chicken meat & egg.

Korver D. R. (2023). Review: Current challenges in poultry nutrition, health, and welfare. *Animal : an international journal of animal bioscience, 17 Suppl 2*, 100755. https://doi.org/10.1016/j.animal.2023.100755

Maharjan, P., Martinez, D. A., Weil, J., Suesuttajit, N., Umberson, C., Mullenix, G., ... & Coon, C. N. (2021). Physiological growth trend of current meat broilers and dietary protein and energy management approaches for sustainable broiler production. *Animal, 15*, 100284.

Sahil Thakur.Impact of rumours and myths about bird's flu outbreak on the poultry industry in social media and strategies to address them in the Indian context. PashudhanPraharee, October 2, 2023

Subashree A. Impact of Poultry Consumption by Adolescent Females - An Analytical Study. Drug Invention Today. 2020;13(6):1220-1223

USDA: United States Department of Agriculture. *Choice Reviews Online.* 2011;48(07):48-3859–48-3859.doi: 10.5860/CHOICE.48 3859.

Xia PF, Pan XF, Chen C, Wang Y, Ye Y, Pan A. Dietary Intakes of Eggs and Cholesterol in Relation to All-Cause and Heart Disease Mortality: A Prospective Cohort Study. Journal of American Heart Association. 2020 May 18;9(10):e015743. doi: 10.1161/JAHA.119.015743. Epub 2020 May 13. PMID: 32400247

INDEX